The Greensboro Blockhouse Project

AN HISTORICAL AND ARCHAEOLOGICAL INVESTIGATION IN GREENSBORO, VERMONT

The Greensboro Blockhouse Project

AN HISTORICAL AND ARCHAEOLOGICAL INVESTIGATION IN GREENSBORO, VERMONT

Jill L. BAKER *and* Patricia L. HASLAM

The Greensboro Blockhouse Project: An Historical and Archaeological Investigation in Greensboro, Vermont

Paperback ISBN 978-1-946824-07-3
Hardback ISBN 978-1-946824-08-0
Ebook ISBN 978-1-946824-09-7

Editing and Interior Design: Janet Angelo
Published in the United States of America by
INDIEGO PUBLISHING LLC

Front Cover: *Background*: map adapted from James Whitelaw, 1790, Northeastern Vermont, from the Whitelaw Manuscript–Northern VT, Harvard Map Collection. *Foreground*: photograph of monument and cows by Nancy Sullivan.

The blockhouse rendering on the title and first chapter pages, drawn by Jill L. Baker, is meant to serve as a visual aid. It does not necessarily depict the Greensboro blockhouse nor is it meant to offer conclusions regarding the architectural features of the superstructure.

Publisher's Cataloging-In-Publication Data:
(Prepared by The Donohue Group, Inc.)

Names: Baker, Jill L., 1964- | Haslam, Patricia Liddle.
Title: The Greensboro Blockhouse Project : an historical and archaeological investigation in Greensboro, Vermont / Jill L. Baker and Patricia L. Haslam.
Description: [Florida] : IndieGo Publishing LLC, [2017] | Includes bibliographical references and index.
Identifiers: LCCN 2017942277 | ISBN 978-1-946824-07-3 (paperback) | ISBN 978-1-946824-08-0 (hardcover) | ISBN 978-1-946824-09-7 (ebook)
Subjects: LCSH: Greensboro (Vt. : Town)--Buildings, structures, etc.--History. | Greensboro (Vt. : Town)--Antiquities. | Fortification--Vermont--Greensboro (Town)--History. | Excavations (Archaeology)--Vermont--Greensboro (Town) | Vermont--History--Revolution, 1775-1783. | BISAC: HISTORY / United States / Revolutionary Period (1775-1800) | ARCHITECTURE / Historic Preservation / General. | SOCIAL SCIENCE / Archaeology.
Classification: LCC F59.G74 B35 2017 (print) | LCC F59.G74 (ebook) | DDC 974.303--dc23

Archaeological Horizons, Inc.
The Greensboro Blockhouse Project

List of Contributors and Participants during the Four Preliminary Seasons of the Greensboro Blockhouse Project

We are deeply grateful to those who helped to make the excavation and publication of the Greensboro Blockhouse Project possible by contributing funds, expertise, time, and hard labor. Your enthusiastic interest and participation in the blockhouse project is very much appreciated.

Financial Donors:

Ms. Gail Abbey

Mrs. and Mr. Rosemary and Charles Averill

Mrs. and Mr. Sarah and Charlie Ayers (in honor of Pat Haslam)

Mrs. and Mr. Mary and Walter Baker

June Bascom and Michael Hoffman

Ms. Joan Bok

Mrs. and Mr. Anne and Michael Cassidy (in memory of Pete Haslam)

Cellars at Jasper Hill

Prof. Allen Davis

The Freeman Foundation

Mrs. and Mr. Pat and Pete Haslam

Mrs. and Mr. Glea and Lawrence Kreider

Mrs. and Mr. Jackie and Rick Lovett

Mrs. and Amb. Susan and Alan Lukens

Mrs. and Mr. Peg and David Marks

Mrs. and Mr. Arthur Newhouse III

Mrs. and Mr. Martha and William Neimi

Mrs. and Mr. Wendy and Robert Parrish

Mrs. and Mr. Gail and Hu Sangree

Mrs. and Mr. Kitty and Clay Simpson

Mrs. and Dr. James Sowles
The Very Rev. & Dr. John Tidy and Jill Baker
Mrs. and Mr. Judy and Bill Whitman
Windham Foundation
Mrs. and Mr. Elizabeth and Paul Wood
Mrs. and Mr. Patricia and Arthur Wright
Mrs. and Mr. Alice and Neal Van Wyck
The Greensboro Historical Society

Donations In-Kind:

Mr. Peter Haslam Jr.
Mrs. Jenny Stoner

The Dig Team:

Mrs. Sarah Ayers
Mr. Walter Baker
Ms. Karen Brancoforte
Mr. Kyle Brancoforte
Ms. Samantha Saliba
The Very Rev. John Tidy
Mr. Bill Atkinson
Mr. Sam Paine
Dr. Jill Baker (Director)

The In-House Dig Team:

Mrs. Mary Baker
Mrs. Pat Haslam (Co-Director)
Mr. Pete Haslam

Special thanks to the property owners, who wish to remain anonymous.

Table of Contents

Acknowledgments

This project was a labor of love, and we are truly grateful to the many people who generously donated their time, talent, and treasure to this endeavor. First and foremost, if it were not for the generosity of the property owners, this project would never have been able to move forward. We are very thankful that they were equally curious to investigate a "pile of rocks" on their property, and allowed us to dig holes on their land for four summers. The dig team received a crash course in archaeological method and theory, and willingly rose before dawn to be in the field by 6.00 a.m., ready to work, which meant removing dense weeds and overgrowth before carefully removing seemingly indistinguishable layers of dirt using picks of varying sizes and trowels.

Equally important was the in-house team (Pat Haslam, M. Baker, and Pete Haslam), who provided much needed support once the dig day ended. Our thanks go to Tim Short and Deborah Holmes for their surveying expertise, which established the excavation grid. We sincerely appreciate the support of the Greensboro community, especially the Greensboro Historical Society, who encouraged us and offered financial support, advice, and delicious dinners.

Although the fieldwork has ended, post-excavation processing continues, which includes publication. We are deeply grateful to Gail Sangree and John Tidy, who read the manuscript and offered constructive comments. We are also deeply grateful to Janet Angelo of IndieGo Publishing for her detailed, diligent, and creative editing. Her thoughtful and encouraging suggestions have made this a much stronger work.

Research of the four blockhouses on the Bayley-Hazen Road was facilitated by Johana Branson of the Peacham Historical Society and her husband, Jock Gill, and Jane Brown of the Cabot Historical Society. We are deeply grateful to them for sharing their knowledge with us. We are grateful to the Greensboro Historical Society for giving us access to their photographic archives.

List of Figures

About the Authors

Pat Haslam (left) and Jill Baker (right) planning the blockhouse manuscript in Greensboro, Vermont, in 2016. Photograph by Kate Haslam Paine.

Jill L. Baker, PhD

Jill is an Independent Researcher in Ancient Near Eastern Archaeology and a Faculty Fellow (adjunct) in the Honors College at Florida International University, Miami, Florida. Jill holds a BA in Biblical Studies from Gordon College, Wenham, Massachusetts, a MA from Gordon-Conwell Theological Seminary, Hamilton, Massachusetts, and a PhD from Brown University, Providence, Rhode Island. From 2003 to 2006 Jill held fellowships at the W. F. Albright Institute of Archaeological Research in Jerusalem, Israel. Her excavation experience includes Harvard University's Leon Levy Expedition to Ashkelon (1990–2004), and the Gesher Excavation (2003–2004) and Tel Zahara Excavation (2006–2009). Jill's post-excavation experience includes the Tel Miqne Publication Project, Jerusalem, Israel, and she taught at the University of Miami from 2006 to 2009 in the Religious Studies Department. From 2010 to the present, Jill has been the Executive Director of Archaeological Horizons,

Inc., a non-profit organization created to facilitate archaeological research. She directed the excavations of the Greensboro Blockhouse Project in Greensboro, Vermont.

Patricia L. Haslam

A native Michigander, Pat's interest in history began in the fourth grade and continues today. She graduated Michigan State University with a BS degree in Decorative Arts in 1953. While raising a family and living in Montpelier, Vermont, she was connected with a great-aunt who provided her with a lengthy letter that detailed family history going back to Connecticut and Rhode Island in colonial days. That led to taking two library science graduate courses and starting a genealogical research business. She became a Certified Genealogist after being accredited by the Board of Certification of Genealogists in Washington, DC in 1973, and continues as a retired associate today. Pat taught beginner and intermediate classes while serving as a library trustee in Stowe, Vermont.

Pat has written and published ten books. These include three award winners: *Three Score & Ten Union Society Being Autobiographical Accounts of the Experiences By Some Early Residents of Stowe, Vermont 1874-1875* (1993), *The Annotated Cemetery Book, Stowe, Vermont 1798-1998: Histories & Inscriptions* (1998), and *Ski Pioneers of Stowe, VT: The First Twenty-Five Years* (2013), as recognized by the International Ski Heritage Association. Pat has written and published eighty-six articles in national and regional genealogical journals, and the Greensboro (Vermont) Historical Society has published many of her articles in its annual journal.

Preface

My archaeological career began in 1990 as a volunteer with the Leon Levy Expedition to Ashkelon, sponsored by Harvard University. Not only was I bitten by the archaeology bug but also the travel bug. From then on I spent the summer months as a staff member for various expeditions, traveling around the Mediterranean basin while completing a doctoral degree at Brown University at the Center for Old World Archaeology and Art, and eventually receiving fellowships that allowed me to conduct research in Jerusalem. In 2010 the excavation with which I was scheduled to work was canceled.

In the spring of that year, my aunt, Pat Haslam, suggested we combine our skills — she as an historian and genealogist, and me as an archaeologist — to find the fabled Revolutionary War period blockhouse in Greensboro, Vermont.

We conducted a surface survey in July of 2010, and based on the architectural remains and artifacts visible on the surface, we decided that further excavation was necessary to determine the true nature of the structure. The property owners generously granted access to the site, and thus began a project that lasted five years and included three more preliminary excavation seasons. In response to this project we formed a not-for-profit organization, Archaeological Horizons, Inc., to raise funds to support the Greensboro Blockhouse Project and other archaeologically related projects.

For me, this project was successful professionally and personally. Professionally, my interests have centered on transitional moments in ancient history when people discovered how to do a certain thing, or when major events altered people's lives, whether those events were societal, cultural, meteorological, or geographical. My main foci have been Bronze Age burial practices and technology in the ancient Near East, but the knowledge I had gained from past research was relevant to the Greensboro blockhouse dig, even though that structure was built in a much more recent era. As will be seen, numerous ancient concepts and techniques are evident in the construction and technology of the blockhouse. Additionally, this project brings to light a lesser-known moment in history: the construction of a military road to support troops and a campaign into southern Canada during the Revolutionary War. Although some consider this to have been a failed military endeavor, the tactics, energy expenditure, and resulting settlement suggest it was quite successful and achieved some of its intended goals.

The Greensboro Blockhouse Project was a sort of micro-dig that operated on a shoestring budget. We excavated for one week per year with a very small team consisting of four or five people digging and three people who supported their efforts, all members of the Baker-Haslam clan. As with most digs, participants develop bonds that last a lifetime. This dig strengthened our bonds as a family as we brought to light nearly a lifetime of research conducted by my Aunt Pat. Although I had known that Pat was actively researching and writing, I had no idea her research was as extensive and prolific as it is. Since the 1970s, Pat has produced at least one article per year and multiple books, something many tenured professors at universities have not done. However, unlike many tenured professors, Pat has enthusiastically and diligently researched and published without the support of a university department or funding. Her work has been out of interest, curiosity, and a desire to make meaningful contributions. Pat Haslam is a true scholar, and I am grateful to have had the chance to share this project with her.

The Greensboro blockhouse archaeological dig was also a time for Pat's husband Pete to see firsthand some of what Pat had been working on for so many years, and to see her research come to fruition. Pete's curiosity was piqued, as evidenced by the questions he asked over our long family dinners. We did not know it at the time, but Pete's body was growing tumors that would ultimately take his life in April 2016. If not for this project, my family and I would have missed out on this valuable time with him. This project also allowed my family to finally understand, on a smaller scale, what I had been doing in Israel all those years spent digging in the dirt. Archaeological excavation is physically demanding and mentally challenging, and now they understood dig life better! I am grateful to have shared this project and experience with them.

It is my hope that this book will combine what is already known about the Bayley-Hazen Road blockhouses, as well as those who built them and resided in them, with new information produced from our archaeological investigation of the site we consider the location of the Greensboro blockhouse based on the evidence we gleaned. Future investigations may prove otherwise, and discussions to the contrary are always welcome. However, with the publication of this book, we hope to share with the wider community the evidence we discovered at the site and the importance of that evidence in the broader scope of American Revolutionary War history.

~ *Jill Baker*

Historical Timeline of the Bayley-Hazen Road in Greensboro, Vermont

April 19, 1775 ~ Battle at Lexington & Concord, Massachusetts.

1775 ~ Gen. Jacob Bayley of Newbury wrote to Gen. George Washington about constructing a shorter route to St. John's, Canada.

1775–6 ~ Benedict Arnold invades Canada.

April 29, 1776 ~ General Washington wrote to General Bayley from New York instructing him to begin construction of the road as soon as possible. Congress approved this order two weeks later.

Early Summer 1776 ~ Road was constructed from Wells River through Peacham to Cabot Plain.

Mid-June 1776 ~ General Washington wrote General Bayley that since his army had retreated from Canada and suffered misfortunes, he was obliged to abandon road construction.

July 1, 1776 ~ Congress issued the Declaration of Independence.

Summer 1778 ~ Plans for another Canadian campaign discussed.

April 1779 ~ In 1779, General Washington gave orders to Colonel Hazen to restart the building of the road.

1779 ~ Cabot, Walden, and Greensboro blockhouses built. Road extended to Hazen's Notch where it ended.

The Greensboro Blockhouse Project

AN HISTORICAL AND ARCHAEOLOGICAL INVESTIGATION IN GREENSBORO, VERMONT

Chapter 1

Historical Background

In April 1776, during the course of the American Revolutionary War, Gen. George Washington issued orders to Col. Jacob Bayley of Newbury, Vermont, to construct a military road that would extend from Wells River, Vermont, into St. John's in Lower Canada.[1] It was Washington's assertion that a quick and efficient route was required to facilitate the movement of troops into southern Canada to mount an offensive from that quarter in the event of a British invasion.

Gen. Washington to Jacob Bayley, New York, April 29, 1776:

> Sir: I received your Favour of the 29th Inst., with Mr. Metcalf's Plan, and Captain Johnson's Journal of the Route from Newbury to St. John's. The Representation that was transmitted to me by the Hands of Colonel Little,[67] I had sent to Congress. Mr. Weatherspoon has been since sent to examine or explore a Route; but, I hear, he is still at Cohoos. The Time of the Congress is so taken up with many Objects of Consequence, that it is impossible for them to attend to every Thing, and as it is of Importance, that every Communication with Canada [Note 67: Col. Moses Little.] should be made as free as possible, it is my Opinion and Desire, that you set about the Road you propose as soon as possible. As you must be the best Judge who to employ you will please to take the whole upon yourself. We cannot at this Time spare Soldiers. You must, therefore, engage such Men as you know will do the Business faithfully, and well. As to their Wages, you must agree with

[1] Wood 1919:264; Bogart 1981:43–47; Duffy 2003:51; Thompson 1842:68; George Washington Papers, Library of Congress, 1741–1799 (http://memory.loc.gov/ammem/gwhtml/.

> them on the most reasonable Terms, and I doubt not that you will in this and every other Instance, serve your Country with Integrity, Honour, and Justice. As you go on, you will upon every Opportunity keep me advised and I will provide for the Expence, which you will be careful in making as light as possible. I am, sir, etc.
>
> P.S. I send you by Mr. William Wallace £250 lawful Money to begin with.[2]

The Continental Congress voted on May 10, 1776, "that as the road recommended by General Washington between the towns of Newbury, on Connecticut River, and the province of Canada, will facilitate the march and return of the troops in that quarter, and promote the public service, the gen'l be directed to prosecute the plan he has informed respecting said road."[3] It was thought that the proposed road would provide a shorter route to Canada, saving some ten days' journey time, as opposed to traveling via Lake Champlain, the traditional route into southern Canada.[4]

The course of the military road was planned by Col. Thomas Johnson "for the passage of men and teams during the construction of the road,"[5] and it was "surveyed by James Whitelaw of Ryegate, and the actual work of the road building was carried out by Gen. Bayley and a party of laborers,"[6] with the help of a scout known as "Indian Joe" and later by Col. Moses Hazen. In July 1776 the course had been marked as far as St. John's, and the road itself constructed as far north as Peacham, allowing the passage of wagons.[7] However, soon after construction began, news of defeat and the driving out of the American army from Quebec by the British as well as potential invasion from the north forced "a hasty abandonment of the undertaking…."[8] The project was suspended for two years, and the road-construction party retired to Newbury. The cost of building this portion of the road in 1776 was £982,[9] which included 110 men employed at a rate of £10 per month for 45 days, plus food and half a pint of rum per day.[10]

On September 11, 1778, General Bayley received orders from General Washington to resume construction of the military road to Canada:

---

[2] Letter from George Washington to Jacob Bayley, April 29, 1776 from the George Washington Papers at the Library of Congress, 1741–1799: Series 2 Letterbooks, Letterbook 9. Web pages: http://memory.loc.gov/cgi-bin/query/P?mgw:6:./temp/~ammem_5UWC::http://memory.loc.gov/cgi-bin/query/r?ammem/mgw:@field (DOCID+@lit(gw040436)).
[3] Bogart 1948:43. Also Crockett 1921:541–546.
[4] Bogart 1948:43.
[5] Bogart 1948:43.
[6] Bogart 1948:43.
[7] Bogart 1948:43. Also *Reminiscences of Jonathan Elkins*, 1921:189.
[8] Wood 1919:264.
[9] At today's rates (2016), £982 equals $1,418.98.
[10] Bogart 1948:43.

> If you find a favorable report from credible people, on the matters herein mentioned, and as your situation is so distant from hence, you may in the month of Novr. next employ a part of Colo. Bedels regiment, should it be continued, or a small number of other good men, in cutting a road from your House[29] [Note 29: Bayley lived at Coos, N.H.] into Canada, which you with others have reported to me to be practicable. Your reasonable expences in this service will be allowed you. You will from time to time transmit to me an account of your proceedings with all the intelligence you shall collect.[11]

However, Bayley's involvement in the project ended at Peacham, Vermont, when Col. Moses Hazen, an officer in the Continental Army, assumed command with orders to complete the project on March 6, 1779.[12]

On March 6, 1779, General Washington sent the following dispatch to Moses Hazen at the army's headquarters in Middlebrook:

> Sir: Immediately upon receipt hereof you are to proceed with your Regiment to Coos. You are to march in three divisions for the benefit of covering your Men, at Night. You will see the whole put in motion before you leave your present Incampment [sic], and will appoint one or more Officers, as the case may require, to collect your straglers [sic] and bring up your rear. Let your Rout [sic] be properly marked and allow a day between each division, the Officers commanding each to be responsible for the good order and conduct of their Men upon the March.
>
> Upon your arrival at Newbury, you will know of Genl. Bayley what Plan he has on foot for intelligence from Canada; and take such measures to obtain fresh advices as seem best adapted to the end. To know the pres't disposition of the force in Canada, and how it will probably be employed in the Spring are essential objects; Intelligence on these points to be gained, if possible, and communicated without delay, to me. While this is doing your Regiment may be employed in extending the Road towards the River Sorrel, or if that shall be deemed too hazardous (till a greater force may be assembled) you may mend and repair what has been already opened by Colo. Biddle.[68]
>
> [Note 68: Col. Timothy Bedel.]
>
> You must not suffer the Officers to keep more Horses than are absolutely necessary for the discharge of their respective duties, as care and saving in the article of Forage and Provisions must be attended to with scrupulous exactness.

---

[11] Letter from George Washington to Jacob Bayley, September 11, 1778: Series 3b Varick Transcripts. The George Washington Papers at the Library of Congress, 1741–1799: Letterbook 6. Web pages:
http://memory.loc.gov/cgibin/query/P?mgw:3:./temp/~ammem_5UWC::http://memory.loc.gov/cgi-bin/query/r?ammem/mgw:@field (DOCID+@lit(gw120451)).

[12] Bogart 1948:46.

> On your March, but in a more pointed manner when you approach Newbury inform yourself with some degree of certainty whether the Inhabitants would give much aid, by their personal Services, in an Expedition by the way of Co'os against Canada, if they could have a well grounded hope of a French fleet and Army appearing in the St. Lawrence to co-operate with them. The result of these enquiries you will communicate to me as soon as you have obtained the requisite knowledge.[13]

Hazen's construction of the road was also prematurely terminated in late August/September 1779, in Westfield, Vermont, known today as Hazen's Notch,[14] when news was received of another invasion from the north. This 54-mile road became known as the Bayley-Hazen Road. For defensive purposes and to create a military presence in the area, four blockhouses were erected at Peacham, Cabot, Walden, and Greensboro, and "garrisons were stationed in the blockhouses along this road from time to time, as occasion demanded, until the end of the war."[15] Of these four blockhouses, the Greensboro blockhouse is the focus of this book.

In its entirety, the Bayley-Hazen Road originated in Newbury and passed through Deweysburgh (later divided into Danville and Peacham), Cabot, Walden, Hardwick, Greensboro, Mindon (Craftsbury), Lutterloch (Albany), and Kelleyvale (Lowell), terminating in Westfield at Hazen's Notch.[16] The road was protected by a series of blockhouses as referenced in a letter from George Washington to Moses Hazen dated July 20, 1779:

> Sir: I have duly received your favors of the 10th. Ult and 10th. Inst. I am obliged to you for the Intelligence from Canada and should be happy if circumstances would authorize an implicit credit to be given to the whole of it. The persons who gave it to Major Whitcomb[75] and Captain Paulant,[76] I fear, have taken it up in several parts upon slender grounds. [Note 75: Maj. Benjamin Whetcomb, of the New Hampshire Rangers. He was retired in January, 1781.] [Note 76: Capt. Antoine Paulint (Paulant), of the Second Canadian Regiment. He was retired in July, 1782.]
>
> The pressing situation of Affairs will not permit me to go into a minute consideration of your Letters; but I am to desire in the most explicit terms that you will not put the public to any expence in those points. I

---

[13] Letter from George Washington to Moses Hazen, March 6, 1779, from the George Washington Papers at the Library of Congress, 1741–1799: Series 3b Varick Transcripts, Letterbook 8.
Web pages: http://memory.loc.gov/cgi-bin/query/P?mgw:7:./temp/~ammem_1SoB::http://memory.loc.gov/cgi-bin/query/r?ammem/mgw:@field (DOCID+@lit (gw140200)).

[14] Baldwin 1906:310; Bogart 1948:47; Miller and Wells 1913:67–78; Duffy et al. 2003:51; Thompson 1842:68–69.

[15] Baldwin 1919:304–305. See also Crockett 1921:261; Wood 1981:48–49; Miller and Wells 1913:67–78.

[16] Baldwin 1919; Thompson 1842:68–69.

have no objection to your *building Block Houses and Stores*, if it can be done entirely by your own people. Your command was to answer a particular Object, intimately connected with or at least intended to promote and facilitate the execution of a plan which I had in view. There cannot be a full communication of the real objects of every command to the Officer detached, and he should always in such cases make his instructions as nearly as possible the rule of his Action. In the present instance, I wish you may not greatly have exceeded my intentions in many things and incurred an expence that will greatly disatisfy [sic] the public. Your Return shall be transmitted to the Board of War. I have granted Warrants to Captain White[77] for the pay of Your Regiment for A and M[78] and he has received the Money. Capn. White will inform you of our success against Stoney point. I am, etc.[79]

[Note 77: Capt. Moses White, of the Second Canadian Regiment. He was aide to Gen. Moses Hazen from September 1781 to the close of the war.] [Note 78: For the months of April and May.] [Note 79: The draft is in the writing of Robert Hanson Harrison.] [17]

Blockhouses were constructed in Peacham, Cabot, and Walden, where Major Walden and several soldiers were stationed during the winter of 1799–1780 and after whom the town was named when chartered, and in Greensboro.[18]

According to the account of Jonathan Elkins, Col. Moses Hazen arrived in Peacham in the year 1779 with 150 men. Hazen and his company were "… cutting the road, building bridges, &c. when they had got 6 miles above Elkins house they built a Log house wit [sic] port holes which was called the first block house … the 2d was bult [sic] in Walden 12 miles above Elkins".[19] The Walden blockhouse was a large log structure measuring approximately 20 by 40 feet. It contained a stone chimney in the middle with 6-foot timber beams running all around the house covered with exterior hewn timbers that elevated the house another six feet and contained portholes, or gun holes, for small arms.

A stockade was erected some 3 or 4 rods[20] from the house, typically 8 or 10 feet in height and made from tree trunks, with one end sunk into the ground and the up-end sharpened. The tops and limbs of trees were sharpened to points, and sharpened brush was prepared in case it was necessary to block the road. Some six or eight miles from the Walden

---

[17] George Washington to Moses Hazen, July 20, 1779, in the George Washington Papers at the Library of Congress, 1741–1799: Series 3b Varick Transcripts, Letterbook 9. Web pages: http://memory.loc.gov/cgi-bin/query/P?mgw:8:./temp/~ammem_OCVl::http://memory.loc.gov/cgi-bin/query/r?ammem/mgw:@field(DOCID+@lit(gw150477)).
[18] Baldwin 1906, Crockett 1921:261.
[19] Elkins *Reminiscence* 1921.
[20] One rod equals approximately 16.5 feet, or 5.5 yards, or 5.0292 meters.

house a third blockhouse was erected in Greensboro, which was similar in its construction to the first one built on the Bayley-Hazen Road.[21]

Bogart describes the "character of the [Bayley-Hazen Road] blockhouses which were erected at various points for purposes of defense."[22] They were "rude structures of logs, hewn on at least two sides, pinned at the corners and perhaps at intervals along their length," and the undergrowth was cleared for some distance from the structures "so that an approaching enemy could not conceal himself."[23]

The blockhouses contained "one strong door" for entry and exit and one "narrow high window" to allow light to enter the building.[24] At about shoulder height, narrow openings were cut all around the structure to function as "rifle port-holes."[25] From the exterior, these appeared as "narrow upright slits" which were only wide enough for the "muzzle of a rifle."[26] However, inside the blockhouse, the log was cut away at a slant to provide a range of movement for the riflemen. There could be as many as ten portholes on the first floor, with the same design on the second floor, which was cantilevered over the first and contained gun loops. Where possible, a deep enough well was dug in the cellar to reach water. This allowed a garrison to withstand a siege without needing to replenish water supplies. Blockhouses such as these could house as many as fifty men.[27]

Troops were posted in the Bayley-Hazen Road blockhouses "... at intervals while the war lasted."[28] In the summer of 1781, Captain Nehemiah Lovewell and his Company were stationed at the blockhouse in Peacham.[29] In September of that year, Lovewell sent four scouts to take control of the blockhouse in Greensboro. On August 10, 1781, while a short distance away from the safety of the fort, the scouts were attacked by Indians.[30] Two were taken captive, Privates Nahum Powers and Nathaniel Martin, and two were killed, Sergeant Constant Bliss and Private Moses Sleeper. The bodies of Bliss and Sleeper remained unburied until Powers and Martin were released from captivity and returned to Peacham. At that time, the decomposed bodies of Bliss and Sleeper were finally buried.

---

[21] Elkins 1921:192.
[22] Bogart 1948:49.
[23] Bogart 1948:49.
[24] Bogart 1948:49.
[25] Bogart 1948:49.
[26] Bogart 1948:49.
[27] Bogart 1948:50.
[28] Miller and Wells 1913:72.
[29] Miller and Wells 1913:72–73; Crockett 1921:266.
[30] Which Native American tribe conducted the raid and took Powers and Martin captive remains a matter of discussion. However, the Abenaki, St. Frances, and Cagnawagah tribes were local to the area (Weber 1990; Elkins 1921).

Following the cessation of hostilities and the declaration of peace in 1783, the original defensive purpose of the blockhouses changed significantly. Having an efficient design for large numbers of people, they functioned as homes, schools, and gathering places for the early settlers to the area.[31] In the spring of 1789, the Greensboro blockhouse became the home of Ashbel and Aaron Shepard and their families, who were early Greensboro residents.[32] Aaron Shepard lived on the property until his death in 1811.[33] Several years later, Aaron Shepard's widow, Phebe Shepard, and Thomas Durkee, administrators of the Aaron Shepard estate in 1813, sold the land and buildings to Joel Cushing.[34] Although there is no documentary evidence, it is reasonable to assume that the blockhouse fell into disrepair and collapsed, perhaps around the time when Shepard and Durkee sold the property to Joel Cushing. Around the same time, Shepard may have started to build the Stone House a short distance to the south, the construction of which Joel Cushing finished.[35] Hemenway documented the location of the Greensboro blockhouse as being "... on the western side of the Caspian, on what was for many years known as the Cushing, and more recently as the William's farm,"[36] and after that the Miller farm.[37] As time passed, the legend of the blockhouse and the fate of Sleeper and Bliss remained part of local lore; however, while the celebrated events were passed down through oral tradition and recorded in local history, the location of the blockhouse was lost to more recent generations.

The site proposed as that of the Greensboro blockhouse resides on private property. Therefore, at the request of the property owners, its exact location will remain confidential, except to say that it is in the vicinity of the monument that was erected in 1941 to commemorate the deaths of Sleeper and Bliss, next to the Bayley-Hazen road in Greensboro, Vermont. Based on the evidence presented throughout this work, it is our belief that this is the location of the legendary blockhouse until convincing evidence can be presented to the contrary.

It would seem that after the Shepards sold the blockhouse and it fell into ruin, several subsequent owners of that property were aware of its existence and significance and never cleared a substantial portion of land around it for farming. The land on which the blockhouse stood appears to have been separated from farmed land by posts and thin wire. It is due to this respectful curation by successive owners that the site was so well

---

[31] Baldwin 1906; Elkins 1921.
[32] Baldwin 1906:306; Haslam 1990:52, 1996.
[33] Haslam 1990:52.
[34] Greensboro Land Records, Vol. A:451, reconstructed due to the loss of all town records and deeds to fire of 1832.
[35] Ca. 1816. Haslam 1996.
[36] Hemenway 1877:210.
[37] *The Vermonter* Vol. II p. 311. See also Haslam 1990:52.

preserved, allowing us to probe into its history and reveal part of its story to the 21st century.

After the erection of the Sleeper-Bliss monument, subsequent generations neglected and forgot the location of the blockhouse, allowing it to slip into legend and folklore, but the construction of the road, its four blockhouses, and the fates of Bliss and Sleeper represent actions and events that actually happened to people who lived and died during the Revolutionary and Post-Revolutionary War periods. Yet to some, these events remain an insignificant piece of the historical puzzle, the mere fascination of local residents, and therefore not worthy of wider mention or serious research.

The reason for this dismissal has to do with the Bayley-Hazen Road. This short 54-mile road, originally intended as a conduit to Canada for military purposes, never performed its intended function. It became militarily obsolete before it was completed, leading some to speculate that its construction was a ruse[38] to occupy the attention of the British forces so that other military maneuvers could be accomplished without hindrance. Regardless of the reason for its construction, the importance of the road cannot be discounted. General Washington, the Continental Congress, Whitelaw, Bayley, Hazen and those who constructed the road committed considerable money, materials, and labor to this project and its cause, which suggests that overall, whether a diversionary tactic or practicality, this project was considered significant by the contemporaneous military leaders.

Once complete, the road was also significant because of the blockhouses constructed at intervals along its length. These blockhouses accommodated garrisons that provided protection for nearby existing communities and newly developing ones. With activity between the British and American forces close by, residents of northern New Hampshire and Vermont were concerned for their safety.[39] The four blockhouses at Peacham, Cabot, Walden, and Greensboro, and the garrisons stationed within, addressed their concerns, both real and imagined.

After the Revolutionary War, the decommissioned blockhouses facilitated the settlement and economy in what remained a relatively untamed, wild portion of north-central Vermont. A road such as the Bayley-Hazen provided an immediate and substantial conduit from Newbury to Westfield (Hazen's Notch) and eventually to Montreal. A well-built road such as this would have been very expensive and taken years for the settlers to build. The focused construction effort of the military resulted in a major north-south travel route that opened the area to settlement.

---

[38] Bogart 1948:46–48; Thompson 1842:68–69.
[39] Elkins 1921; Bogart 1948:43–54.

In the years following the war, numerous towns were established as early settlers found fertile farmland, and products were transported between Montreal and Boston via the Bayley-Hazen Road.[40]

While the Bayley-Hazen Road and its blockhouses may not be as famous as those of Boston, Philadelphia, or Ft. Ticonderoga, they do reflect the northern expansion of the young American country and economy as well as the ambition and military prowess of Gen. George Washington. Like the blockhouses found at Ft. Kent, Maine, Ft. Pitt, Pennsylvania, Ft. Constitution/Ft. Lee, New Jersey, and LaColle Mills, Canada, these solitary sentinels contributed to both the revolutionary cause and the growth and development of the young country, making their role in American's early history unique and valuable. The Greensboro Blockhouse Project brings to light this little known piece of the Revolutionary War puzzle.

[40] Bogart 1948:48.

Chapter 2

Supporting Evidence for the Greensboro Blockhouse Location

Pat Haslam, co-author of this work, had an idea as to the location of the blockhouse. Based on her extensive research, she had learned that earlier inhabitants of Greensboro were apparently aware of the existence of the blockhouse, but its precise location had been lost to recent generations. Nevertheless, careful scrutiny of the available maps and textual references, and the stories passed down through local oral tradition, offered evidence that enabled us to identify its location with a considerable degree of confidence.

Several early maps situate the blockhouse near Caspian Lake (Figure 2.1). Two maps compiled by James Whitelaw, Vermont Surveyor General, one of which was drawn in 1790, clearly marks the location of "Hazen's Blockhouse" on the western side of Caspian Lake and on the western side of the Bayley-Hazen Road.

The second Whitelaw map dated February 1794 also marks the blockhouse on the western shore of "Beautiful Lake." Another by Wilgus compiled in 1791 also depicts "Hazen's Block House" on the western shore of Caspian lake but on the eastern side of the Bayley-Hazen Road (cf. Figure 3.11, Chapter 3).

Finally, a fourth, by Carl Bohn of Germany compiled in 1796, also identifies the blockhouse on Caspian Lake's western shore. Curiously, however, Bohn locates the blockhouse on the opposite side of the Bayley-Hazen Road.

Figure 2.1. *Left*: James Whitelaw, 1790, Northeastern Vermont, adapted from the Whitelaw Manuscript-Northern Vermont. Harvard Map Collection. *Middle*: James Whitelaw, 1794, map of Beautiful Lake (Caspian Lake). Adapted from The History of Greensboro The First Two Hundred Years (1990:16). *Right*: Carl Bohn, 1796, map of Vermont. Adapted from: (http://www.old-maps.com/vt_state/vt_1796_Stotzmann_1m.jpg).

The Whitelaw maps situate the blockhouse immediately west of the Bayley-Hazen Road and describe it as either "Hazen's Blockhouse" on "Hazen's Road" (1790) or "Old Blockhouse" on "Hazen's Road so-called" (1794). The Bohn map (1796), conversely, situates it on the eastern side of the road and describes it as "Scotland Hazen's Blockhaus." Though published only six and two years apart respectively, the discrepancy between these maps has caused some to speculate that the course of the Bayley-Hazen Road may have migrated during that brief period (cf. below, Chapter 3).

Historians of the late 19th and early 20th centuries, specifically Abby Maria Hemenway (1877), Frederick W. Baldwin (1905/1906), E. Miller and F. P. Wells (1913), F. J. Wood (1919), and W. H. Crockett (1921), identified the location of the blockhouse on the western shore of Caspian Lake. Baldwin, in reference to the slain scouts, remarks, "No monument has been erected thus far to their memory, altho [sic] I am informed a flag pole has been erected on the supposed site of the block house, and a boulder beside the highway to the west has been so lettered as to direct those interested to the spot."[41] This was in addition to the aforementioned reference by Hemenway locating the fort on the Cushing and later the Williams farm.[42] P. Haslam's title search confirms that J. Cushing, J. Williams, and I. F. Williams indeed owned said property.[43]

---

[41] Baldwin 1905:311.
[42] Hemenway 1877:210.
[43] Haslam 1990:52.

Although no monument had been erected to the slain scouts when these historians wrote, efforts to do so were underway. In a letter dated 1903 written by Rev. Perrin Fisk to John Bray Cook, Fisk states that a monument should be established to mark the blockhouse ruin and to commemorate the lives of the military scouts Constant Bliss and Moses Sleeper.

In 1941, a memorial to Constant Bliss and Moses Sleeper was erected next to the Bayley-Hazen Road on the eastern side of it. In August 1942, the Report of The Blockhouse Memorial Committee to the Greensboro Association documented the donation of the land and monies that supported the project. The report also described in detail the preparation of the land and placement of the monument. Presumably this monument was situated in the vicinity of the ruined blockhouse by those who remembered its location, but a separate marker was never established for the ruined fort. (Appendix 1, Letter and Report)

Additional support for the identification of the rock pile as the ruined blockhouse comes from Raymond Mercier,[44] who during this conversation, indicated in handwriting on a 1986 map of the Caspian Greenlands Corporation the location of a "cellar hole" and an "old well." (Figure 2.2, next page)

According to the title search, W. and M. Mercier owned the Cushing/Williams property from 1925 to 1961, and Raymond was their son. Mercier's identification of these locations is significant since his parents, Mr. and Mrs. Wilfred Mercier, gifted the above mentioned "...plot of land on which the monument stands...."[45] If Mr. and Mrs. Wilfred Mercier were aware of the location of the blockhouse ruin, it is reasonable to assume that their son was too.

---

[44] Personal communication between Raymond Mercier and Pat Haslam in 1994.

[45] Appendix 1, Report of the Blockhouse Memorial Committee to the Greensboro Association. August 15, 1942.

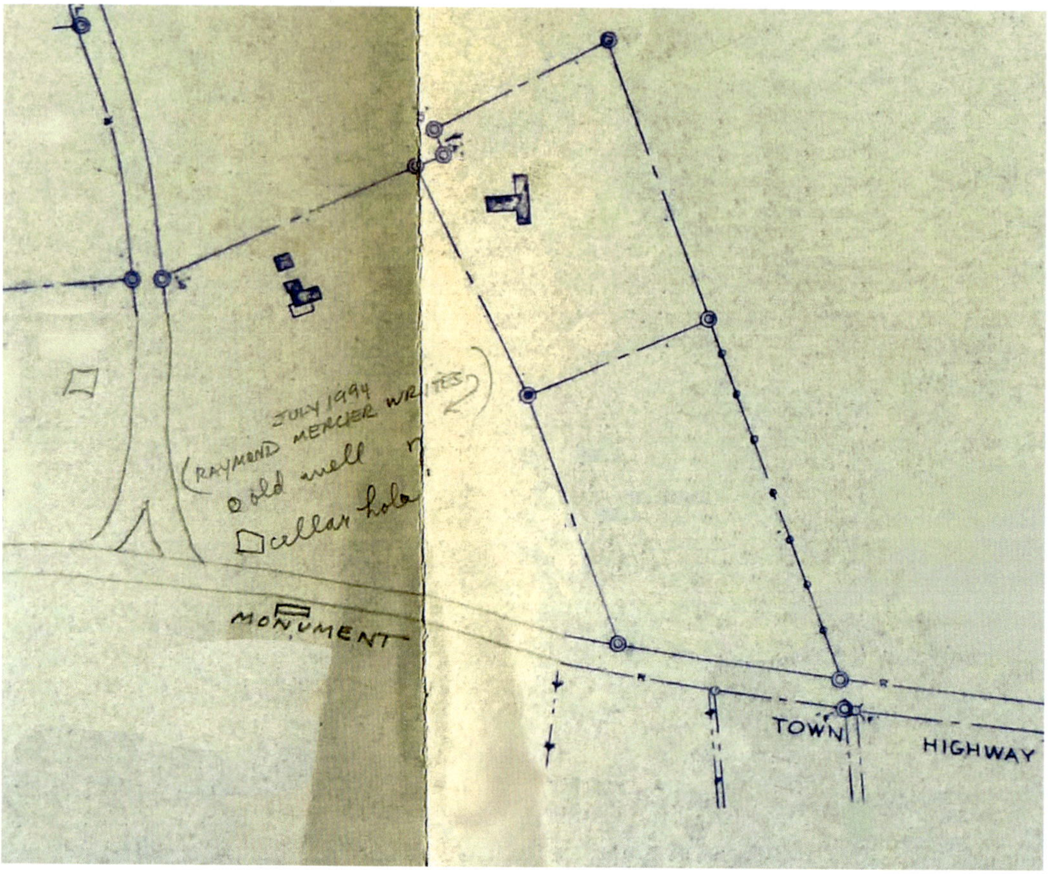

Figure 2.2. R. Mercier indicated in handwriting the location of a "cellar hole" and an "old well" thought to be those of the blockhouse. Additionally, Mercier situated the "monument" on this map. Adapted from the Caspian Greenland Corp. and prepared by R. N. Bohlen in 1986.

Finally, based on the data provided by the 2010, 2011, 2012 and 2015 Preliminary Seasons[46], the area referred to as the "cellar hole" most likely represents a collapsed building that dates to the late 18th and early 19th centuries, and therefore it is reasonable to conclude that it is a Revolutionary War-era blockhouse. The term blockhouse is derived from the German *blokhüs*, Dutch *blokhuis*, and Swedish *blockhus*. The blockhouse was used as a defensive structure in the ancient Near East, Europe, and Britain, and was brought to America by early European and English settlers.[47]

---

[46] It will be noted that the terms 'preliminary season(s)' or 'preliminary excavation season(s)' are used to refer to the work conducted at this site thus far. The reason for qualifying our work as preliminary is because it has been exploratory, probing limited areas of the site, rather than full and complete excavation.

[47] It should be noted that prior to the European invasion, Native American peoples of North America constructed and utilized defensive structures. For example, Mesa Verde,

A blockhouse was a small, isolated, two-story building with one lower and one upper room. It functioned as a strong point and/or a foothold for an advancing military force. The purpose of the blockhouse was to protect a specific area for a precise period, and it was intended to provide temporary accommodation for garrisons as their campaigns moved from one place to another. Blockhouses were also built into fortification walls at specific intervals, whether in the length of the wall or at the corners, in which case they functioned as watchtowers.

The Romans were one of the earliest peoples to use blockhouses in this manner and to call them watchtowers. Examples of Roman watchtowers may be found at Limes Germanicus (ca. 83–260 CE) and Vecht (Fectio), Germany, near the Vecht River (Figure 2.3). In general, the first floor was used to store equipment, food, and armament. The second floor was used as living quarters and for cooking and sleeping. The third floor was for on-duty personnel who kept watch through the day and night.

Figure 2.3. *Left*: Limes Germanicus. Roman watchtower ca. 150 m from Kohorten-Kastell Zugmantel. (Photograph from Wikipedia, "Turm Zugmantel" by No machine-readable author provided. RalfZi assumed (based on copyright claims). - No machine-readable source provided. Own work assumed (based on copyright claims). Licensed under CC BY-SA 3.0 via Commons: https://commons.wikimedia.org/wiki/File:Turm_Zugmantel.jpg#/media/File:Turm_Zugmantel.jpg). *Middle*: Reconstructed Roman *specula* or *vigilarium*, 'watchtower' in Germany. (Photograph from Wikipedia, https://commons.wikimedia.org/wiki/File:Wp12_77_Rekonstruktion.jpg). *Right*: Reconstructed watchtower near Fectio, Vecht, Germany. (Photography from Wikipedia, https://commons.wikimedia.org/wiki/File:Romeinse_wachttoren_fort_vechten_netherlands.jpg).

Colorado; Fort Ancient, Lebanon, Ohio, Cahokia (700–1400 CE) in Illinois, and the Etowah (950–1450 CE) of Illinois. Kaufmann and Kaufmann 2007:10–11.

As the Roman Empire expanded, armies constructed these watchtowers to provide shelter and protection for advancing military units as they conquered villages and cities. When the legion decided to occupy a specific location, they first built an earthen rampart upon which they constructed a tall wooden palisade, or stockade fence, made of hewn logs sunk deep into the ground for stability, their tops carved into sharp points. Behind the rampart and palisade was a ditch, and behind that, a watchtower, which was erected upon an earthen mound as well. Standing two or three stories tall, watchtowers were made of stone, wood, or both, though usually with a stone foundation and wooden superstructure. The roof was pyramidal in shape, and there was often a balcony accessible from one of the upper floors.[48] The Romans built their watchtowers so that each was within sight of the next, which allowed them to use a semaphore system of communication that involved sending visual signals to each other using blades, paddles, mirrors, or fire.[49]

Later, the watchtower became known as a blockhouse, probably due to its square or rectangular shape. An early example of a blockhouse was the Cow Tower in Norwich, England, constructed 1398–1399 (Figure 2.4). The Cow Tower was made of brick and had three floors. The upper floors contained gun portholes, six per floor. Its initial purpose was to provide protection from French and English rebels.

Between 1539 and 1545, King Henry VIII erected a series of blockhouses at the Thames Estuary, Solent, and Plymouth as part of his maritime defense program. These typically consisted of square, rectangular, or circular stone-built towers located at strategic maritime locations to block access to crucial points as well as the advance of enemy troops. These were usually outfitted with a bastion and various other weapons. The Henrican blockhouse, 1545, at Mt. Edgcombe, Plymouth, in Devon, England (Figure 2.4, next page), serves as an excellent example of one such blockhouse.

[48] http://www.figuras.miniatures.de/ancients-2-roman-watchtower.html.
[49] http://www.figuras.miniatures.de/ancients-2-roman-watchtower.html.

Figure 2.4. *Left*: Cow Tower, Norwich, Norfolk, England. Borrowed from Wikipedia. ("Cowtower" by Lavag at English Wikipedia - Transferred from en.wikipedia to Commons. Licensed under GFDL via Commons, https://commons.wikimedia.org/wiki/File:Cowtower.jpg#/media/File:Cowtower.jpg). *Right*: The Henrican Blockhouse, 1545, Mt. Edgcombe, Plymouth, Devon, England. Borrowed from Wikipedia ("Mount Edgcumbe blockhouse" by Nilfanion - Own work. Licensed under CC BY-SA 3.0 via Commons. https://commons.wikimedia.org/wiki/File:Mount_Edgcumbe_blockhouse.jpg#/media/File:Mount_Edgcumbe_blockhouse.jpg.

As English and European explorers and settlers ventured into the New World, blockhouse technology came with them. Although the purpose of the blockhouse was much the same as in the Old World, their shape, layout, and the materials used in their construction varied and was more location specific. In the New World, blockhouses were square or rectangular and made of wood, stone, or whatever local materials could be quickly and easily fashioned into a fortress.[50]

These military structures, as described by Sébastien le Prestre de Vauban (1633–1707) and Lieutenant Henry Yule (1851), were much more solid than the average house or barn of the period. Because they were defensive structures, the blockhouses consisted of walls and foundations that were wider (thicker) than those of a civilian residence.[51] The logs that comprised the walls had to be at least 12 inches thick, and the height of the first floor ceiling at least 9 feet high. The width of the second floor

[50] Yule 1854:49; Morrison 1952:20–79; Wilbur 1992:28–30; Kaufmann and Kaufmann 2004; Isham 2007.

[51] Yule 1854:49; Morrison 1952:20–79; Wilbur 1992:28–30; Kaufmann and Kaufmann 2004; Isham 2007.

extended or cantilevered over the walls of the first floor and contained portholes through which soldiers could deploy ramrods and firearms.[52]

Blockhouses also incorporated gun loops into their defenses, the design of which was based on ancient arrow slits or arrow loops. The invention of the arrow slit is credited to Archimedes during the siege of Syracuse from 214 to 212 BCE. The arrow slit was a thin vertical opening, sometimes also with a thin horizontal aperture that allowed archers and longbowmen to shoot arrows at attackers from a place of safety inside the castle wall. The arrow slit was accessed by a wide interior opening called an embrasure, which allowed the bowmen range of movement and vision. Examples can be seen in the castle fortresses found in the ruins of Belvoir Castle and the Crusader gate at Caesarea Maritima, Israel (Figure 2.5).[53]

Figure 2.5. *Left* and *middle:* Belvoir Castle, Israel. Crusader fortress (ca. 1168) 12 miles (20 km) south of the Sea of Galilee. *Right*: Caesarea, Israel. Crusader gate (12th–13th centuries). Photographs by Jill L. Baker.

Arrow slits were also utilized in Greek and Roman defensive systems, but their design concept was lost and not rediscovered until the end of the 12th century when they were built into the Dover and Framlingham castles in England and Château Gaillard in France, at which point arrow slits became standard defense mechanisms in castle walls and bastions.[54]

Keeping the basic characteristics of a blockhouse in mind — strong, pyramidal roof; substantial foundation; square or rectangular layout with centralized hearth and chimney; often hastily built for defensive purposes — these differed greatly from proper fortresses, garrison houses, and residential dwellings. Several examples include the Lacolle Mille Blockhouse in Ontario, Canada; Fort Halifax, in Winslow, ME; the Clargue Blockhouse in Ontario, Canada; and the Fort Kent Blockhouse, in Fort Kent, ME (Figure 2.6). Blockhouses could either be freestanding,

---

[52] Yule 1854:49; Morrison 1952:20–79; Wilbur 1992:28–30; Kaufmann and Kaufmann 2004; Isham 2007.
[53] Jones and Renn 1982.
[54] Jones and Renn 1982.

one of several buildings, or integrated as a sort of tower into a stockade enclosure wall. For example, the American Blockhouse site at Mount Independence State Historic Site, in Orwell, Vermont, included a storehouse as part of the defensive compound.[55]

Lacolle Mills Blockhouse
Ontario, Canada

Fort Halifax
Winslow, ME

Clergue Blockhouse
Ontario, Canada

Fort Kent Blockhouse
Fort Kent, ME

Figure 2.6. Examples of northeastern American blockhouses (from left to right). Lacolle Mills Blockhouse, built ca. 1781 as part of the British defense network. This picture, circa 1920, is borrowed from many-roads.com (http://www.many-roads.com/2010/07/06/lacolle-battles/). Fort Halifax, built ca. 1754/1755 by the Province of Massachusetts Bay. This picture, taken in 1936, is borrowed from Wikipedia (http://en.wikipedia.org/wiki/Fort_Halifax_%28Maine%29). The Clergue Blockhouse was originally built in the 1790's (?), restored in 1922, and relocated in 1996. This picture is a postcard from 1902 and is borrowed from digitalgallery.nypl.org. http://digitalgallery.nypl.org/nypldigital/dgkeysearchdetail.cfm?trg=1&strucID=130682&imageID=62805&word=Clergue%20Blockhouse&s=1&notword=&d=&c=&f=&k=0&lWord=&lField=&sScope=&sLevel=&sLabel=&sort=&total=1&num=0&imgs=20&pNum=&pos=1). Fort Kent was built ca. 1838–1839. This picture was taken in 1911 and is borrowed from Maine Memory Network (http://www.mainememory.net/artifact/31460/).

The dimensions and composition of the above referenced structures are similar to those that one would expect to find at the Greensboro site. For example, the Fort Kent blockhouse was a two-story structure, and the walls were constructed using square-hewn cedar logs 19 inches or wider. The second story was cantilevered over the first, and there was only one entry on the ground floor. Another example is the Lacolle Mills Blockhouse (Figure 2.6), which was also constructed using square-hewn wooden timbers. Each wall contained small openings that served as portals for gun barrels and cannons. There are two stories, a fireplace and chimney, and a pyramidal roof. Like most of the blockhouses referenced above, Lacolle Mills Blockhouse has been restored and is open to the public. A cut-away model at the site provides a visual example of the blockhouse's construction (Figure 2.7).[56]

---

[55] http://www.hmdb.org/Marker.asp?Marker=19366.

[56] FortWiki, http://www.fortwiki.com/Lacolle_River_Blockhouse; http://www.historicplaces.ca/en/rep-reg/place-lieu.aspx?id=6386; http://www.ileauxnoix.com/eng/tourisme/blockhaus.html.

Figure 2.7. Lacolle Mills Blockhouse, Quebec, Canada. Cut-away revealing the structure's construction. Borrowed from Wikipedia, (http://en.wikipedia.org/wiki/lacolle_Mills_Blockhouse)

The interior layout consisted of one large room on each of the two floors. From this model it is possible to gain a sense of the proportions of the foundation upon which the structure was built as well as its interior layout. It is reasonable to assume and expect that a similarly substantial foundation would have been used for the construction of the Greensboro blockhouse. Additionally, Jonathan Elkins' description of the Peacham, Walden, and Greensboro blockhouses mentions architectural details similar to the examples mentioned above.

As stand-alone defensive structures, blockhouses were usually surrounded by a stockade wall made of timbers at least 9 inches wide positioned vertically into the ground with sharply pointed ends facing upward. In his reminiscence, Elkins described the stockades of the Peacham, Walden, and Greensboro blockhouses at a distance of 3 or 4 rods (49.5 ft.; 16.5 yards; 15 meters) from the blockhouse. Ditches and earthworks, such as ramparts, also surrounded the structure.[57] Earthen ramparts not only added to the stability of the structure (stockade or foundation wall) but were also a vital part of its defensive strategy.

For domestic buildings, foundations were generally dug into the ground, one foot below frost level, to avoid heaving due to the freezing and thawing action of the ground. They also had to extend at least eight inches above the surface of the ground.[58] Presumably, the foundations of

[57] Elkins 1921; Yule 1854.
[58] Wilbur 1992:28.

a blockhouse would have adopted similar specifications. In some instances, the first-floor walls were constructed of stone rather than timber. An example of this is the Clergue Blockhouse (Figure 2.6) located in Sault Ste. Marie, Ontario, Canada. In 1922, the Sault Ste. Marie Historical Society learned about the condition of that blockhouse from Francis H. Clergue when he became owner of it. In his description, only the first-floor stone walls remained. The property was surrounded by boulders, and the stumps of the stockade remained, some of which were still above ground. There was only one large room, which contained a series of six-inch squares, presumably for gun barrels. Although the wooden timbers had long since disappeared, Clergue reconstructed the blockhouse based on the remnant of the first floor stone walls.[59] According to Elkins' description of the three blockhouses on the Bayley-Hazen Military Road, a stockade with similar construction and dimensions was employed.

Blockhouses should not be confused with garrison houses or forts. Garrisons were very solidly built fortified houses. They adopted a regular house plan and resembled a typical house, as exemplified by the William Damme Garrison House in Dover, New Hampshire; the Gilman Garrison House in Exeter, New Hampshire; and the McIntire Garrison House, York, Maine (Figure 2.8).

William Damm Garrison House, Dover, NH — Gilman Garrison House, Exeter, NH — McIntire Garrison House, Scotland, ME

Figure 2.8. Garrison Houses. *Left*: William Damme Garrison House, Dover, New Hampshire. Built in 1675, this picture was taken ca. 1916. This photograph is borrowed from SeacoastNH.com (http://www.seacoastnh.com/woodman/garrison/13.html). *Middle*: Gilman Garrison House, Exeter, New Hampshire. Built ca. 1700, this picture dates to 22 September 1936 and is borrowed from Wikipedia (http://en.wikipedia.org/wiki/file:Gilman_Garrison_House). *Right*: McIntire Garrison House, York, ME. Built ca. 1707, photograph dated 29 April 1936 and is borrowed from Wikipedia (http://en.wikipedia.org/wiki/McIntire_Garrison_House).

[59] Sault History Online, http://www.cityssm.on.ca/library/Clergue_Block.html; FortWiki, http://www.fortwiki.com/Clergue_Blockhouse; City of Sault Ste. Marie, http://saultstemarie.ca/City-Hall/City-Departments/Community-Development-Enterprise-Services/Community-Services/Recreation-and-Culture/Historic-Sites-and-Heritage/Ermatinger-Clergue-National-Historic-Site/Clergue-Blockhouse.aspx; Canada's Historic Places, http://www.historicplaces.ca/en/rep-reg/place-lieu.aspx?id=5595.

However, garrison walls were thicker, their foundations more substantial, and their fireplaces cavernous enough for cooking large-quantity meals. Garrisons provided protection and accommodation for multiple families during times of attack.[60]

By contrast, forts were elaborately constructed and well-organized structures that comprised barracks for troops, earthen ramparts, ditches, places to house and fire cannons and other firearms, towers, and blockhouses, which were often incorporated into their walls. Several examples include Fort Independence, Boston, Massachusetts; Fort Ticonderoga, Fort Ticonderoga, New York; and Fort Crown Point, Crown Point, New York (Figure 2.9).

Fort Independence
Boston, MA

Fort Ticonderoga
Ft. Ticonderoga, NY

Fort Crown Point
Crown Point, NY

Figure 2.9. Fortress. *Left*: Fort Independence, one of the oldest forts in America, was originally built in 1634 to defend the coast from the British. It has a pentagonal shape and five bastions and has undergone several renovations. The picture above represents the eighth renovation dated to 1834–1851. *Middle*: Fort Ticonderoga was built between 1754 and 1757 as a defensive structure against Great Britain and France. *Right*: Fort Crown Point was a British fort built in 1759 by the British and Provincial troops as defense against the French. All three pictures are adapted from Google Earth.

As opposed to stand-alone blockhouses, which were often hastily built, forts were constructed over a longer period with calculated precision. The most frequently adopted layout was the star-shaped plan because it offered more effective defensive capabilities. However, based on factors such as topography and strategic issues, their shape could also be triangular, square, rectangular, hexagonal, or pentagonal.[61]

Domestic dwellings of the colonial period generally followed one of several plans: a one-room plan, a two-room plan, an added lean-to plan, or an original lean-to plan. Examples are the Fairbanks House, Dedham, Massachusetts; the Whipple House in Ipswich, Massachusetts; the House

---

[60] Morrison 1952:77.
[61] Kaufmann and Kaufmann 2007; Mueller 2006.

of Seven Gables in Salam, Massachusetts; and the Corwin House in Salem, Massachusetts (Figure 2.10).[62]

Fairbanks House
Dedham, MA

Whipple House
Ipswich, MA

House of Seven Gables
Salem, MA

Corwin House
Salem, MA

Figure 2.10. The Fairbanks House was built ca. 1637–1641 and is considered the oldest timber-framed building in North America (http://fairbankshouse.org/). The Whipple House was built possibly as early as 1638 by John Fawn, who sold it to John Whipple. The Whipple family occupied the house for six generations and made numerous additions (Ipswich Museum http://ipswichmuseum.drupalgardens.com/). The House of Seven Gables was built in 1668 by John Turner and occupied for three Turner generations (The House of Seven Gables http://www.7gables.org/index.htm). The Corwin House was purchased by Jonathan Corwin in 1675 as a partially built, timber-frame structure (Historic Buildings of Massachusetts http://mass.historicbuildingsct.com/?p=103; Salemweb: http://www.salemweb.com/witchhouse/).

The foundations of domestic homes were generally constructed of fieldstone[63] and extended at least eight inches aboveground.[64] Based on published plans, it appears that the foundations of houses could be up to eighteen inches wide.[65] In New England, foundations were usually dug below ground level to make room for a cellar, which was generally used to store perishable foodstuffs. If deep enough, and with a separate entryway, animals may also have been housed in the cellar.[66] The sill was laid upon the level surface of the stone and was made of large, heavy, hewn timber beams that carried the posts of the frame, floor joists, and wall studs.[67]

In summary, based on the textual, cartographical, and oral evidence, we feel confident in identifying our dig site as the probable location of the blockhouse in Greensboro, Vermont. Further supporting evidence comes from architectural features and material culture, the stuff that people leave behind. In light of what we know about the construction and layout of a typical blockhouse from military specifications and

---

[62] Morrison 1952:20–22; Isham 2007.
[63] Morrison 1952:24; Wilbur 1992:28; Cummings 2002; Donnelly 2003:34–45; Thorson 2005:82–84, 110–111; Isham 2007.
[64] Wilbur 1995:28.
[65] Isham and Brown 1900; Isham 2007.
[66] Isham and Brown 1900; Morrison 1952:25; Cummings 2003; Donnelly 2003:36; Thorson 2005:110–111.
[67] Morrison 1952:25–26; Wilbur 1992: 28; Donnelly 2003:36; Thorson 2005:110–111; Isham 2007:23, 137.

contemporaneous personal accounts, the design of the collapsed building at the site reveals the structure's purpose.

For the sake of comparison, the preceding discussion briefly described not only the characteristics of a blockhouse but also those of a domestic dwelling, a garrison house, and a fort. The design and construction of a blockhouse is very different from that of the other structures, and therefore is recognizable as such in the archeological record. The chapters that follow will further examine the character and purpose of the collapsed ruin based on the results of archaeological investigation. However, before we delve into the archaeological data, we will discuss the environment and climate of the area in which the Bayley-Hazen Road blockhouses are located, and examine the lives and characteristics of those who forged the road, built the blockhouses, and established the earliest settlement and surrounding neighborhoods.

Chapter 3

Four Blockhouses on the Bayley-Hazen Military Road 1776 & 1779

Many articles have been written about the Bayley-Hazen Military Road over the years: including F. Baldwin's 1906 article "History of the Hazen Military Road" in *The Vermonter* (Vol. XI, No. 16, pages 297–323); Marcus McCorison and Van H. English's invaluable *Atlas of the Bayley-Hazen Military Road 1776 & 1779* (1959); and Crane Brinton's article "The Hazen Road" in *Vermont Life Magazine*, Spring 1955:32. However, few have discussed the details of the blockhouses themselves.

The most comprehensive contemporaneous account of the building of the road and the blockhouses was described in the "Reminiscences of Jonathan Elkins," reprinted in *The Upper Connecticut*, Vol. III, 1943:270–271, the original manuscript of which is housed in the collections of the Vermont Historical Society. This work is invaluable because Jonathan Elkins Jr. (1761–1852) was on the scene. He was taken captive by the British in his father's house, marched to Quebec in 1781, and transported to Ireland and England, after which he was exchanged before being returned to America in 1782.

Concerning the construction of the road, Elkins writes, "In June 1779 Col. Moses Hazen came on to Peacham with about 150 of his men and encamped near our house, the remainder of his regiment was kept at Haverhill, New Hampshire, and they begun [sic] cutting the road, bulding [sic] bridges, &c when thay [sic] had got 6 miles above Elkins house, thay bult [sic] a Log house with port holes which was called the first block house."[68]

P. B. Fisk in his Decoration Day Address in Greensboro (1901) wrote, "... the character of the blockhouses which were erected at various points for purposes of defense . . . were rude structures of logs, hewn on at least two sides, pinned at the corners and perhaps at intervals along their length. One strong door afforded ingress and egress, and a narrow high window admitted light. Narrow openings were cut through a log all the way around, about shoulder-high, to serve as rifle portholes. From the outside these would appear as narrow upright slits, just wide enough to pass the muzzle of a rifle; but inside the log was cut away so as to give the riflemen space to swing his rifle a foot to right or left. There would be perhaps ten such portholes around the lower story.

The upper story was a repetition of the lower, but was built out beyond this, and the portholes were cut downward to prevent an attacking enemy from setting fire to the house. A steep roof, made of poles and bark, usually covered the structure. The cellar was ordinarily dug with, if possible, a spring of water to enable the garrison to withstand a siege. As many as 50 men could find refuge in a typical blockhouse."

Several other blockhouses and forts were strategically placed in towns in eastern Vermont along the Connecticut River to provide protection against the Native Americans and the British prior to 1730 to the end of the Revolutionary War. One such fort was at Newbury, which acted as a hub and base for the early stages of the Bayley-Hazen road: "The Newbury blockhouse (1777–1782) was a large militia log blockhouse with defensive ditch, built for the protection of the town settlers against the British and Indians from Canada. It could shelter up to two militia companies and was located on the ridge north of the town cemetery at the River Oxbow. Gen. Moses Hazen's Continental troops may have used the blockhouse in 1779 as they stored supplies and constructed the Hazen Road system. Located at the south end of the village was Col. Robert Johnston's stockade, a palisaded structure for defense. It was later transformed into a storage barn."[69] Several other minor forts worth mentioning include one along the upper Connecticut River at Fort Dummer (1724–1763), a small fort and a powder house at Corinth, and Stephen's Fort at East Ryegate (1777–1782). Fort No. 4 at Charlestown, New Hampshire, on the east side of the Connecticut River was an important stopping place for travelers making their way through the wilderness of the New Hampshire-Vermont area from 1735 onward.[70]

---

[68] Elkins: 1921:192–193.
[69] Northeast American Forts, northeastamericanforts.com/East/vt.html; Wells 1902:597.
[70] Northeast American Forts, northeastamericanforts.com/East/vt.html.

By studying contemporaneous maps of the Bayley-Hazen Road between 1776 and 1800, one can ascertain whether any blockhouses were drawn on these maps, which might provide a *terminus ante quem*. These included the Abel Sawyer map (1784),[71] the Blodgett/Amos Doolittle map (1789),[72] the James Whitelaw map (1790),[73] the Carey map (1795),[74] the J. Whitelaw map (1794),[75] the Sotzman/Bohn map (1796),[76] and "A Map of Vermont as it was about 1791" by William J. Wilgus (1941).

Interestingly, of these maps, the only Bayley-Hazen Road blockhouse to be noted was the one in Greensboro, but only on some of the maps, including the 1790 and 1794 Whitelaw maps, the 1796 Bohn map (Figure 2.1 above), and the Wilgus map of 1791 (Figure 3.11), described as "Hazens Block House". The Walden, Cabot, and Peacham blockhouses were not shown on these maps. The Greensboro blockhouse was noted on Whitelaw's 1790 and 1794 maps as "Hazen's Blockhouse" and "old blockhouse," respectively, and Bohn's 1796 map as "Scotland Hazen's Blockhaus" (See Figure 2.1). The name "Scotland" may have derived from the Scottish settlements on the Connecticut River and surveyor James Whitelaw's connection to that area.

If the only blockhouse noted on these maps is the one in Greensboro, how should this be interpreted? The most likely scenario is that the Peacham blockhouse was already abandoned or in ruin, and the Cabot and Walden blockhouses were either in ruin or no longer housed the newest settlers in town. We also know that the first settlers to Greensboro settled in that blockhouse in the spring of 1789, and by at least 1816, the nearby Stone House was in the process of being built.[77] (Aaron Shepard may have started work on the Stone House before his death in 1811.) William J. Wilgus' map,[78] "The Role of Transportation in the Development of Vermont As It Was in 1791", depicts no forts along the Bayley-Hazen Road, none at Peacham, Cabot, and Walden, and only one blockhouse symbol on the east side of the road along the western bank of Caspian Lake, which he noted as "Hazen's Block House."

Measuring the distance in miles per the scale from Walden (if the scale of Wilgus' map is accurate), this blockhouse would probably be at Greensboro, and thus the most recent of the four blockhouses.

---

[71] Cobb 1971:53.
[72] Cobb 1971:68.
[73] Cobb 1971:84.
[74] Cobb 1971:118.
[75] Cobb 1971:122.
[76] Cobb 1971:124
[77] Haslam 1996:16.
[78] Wilgus 1791:49.

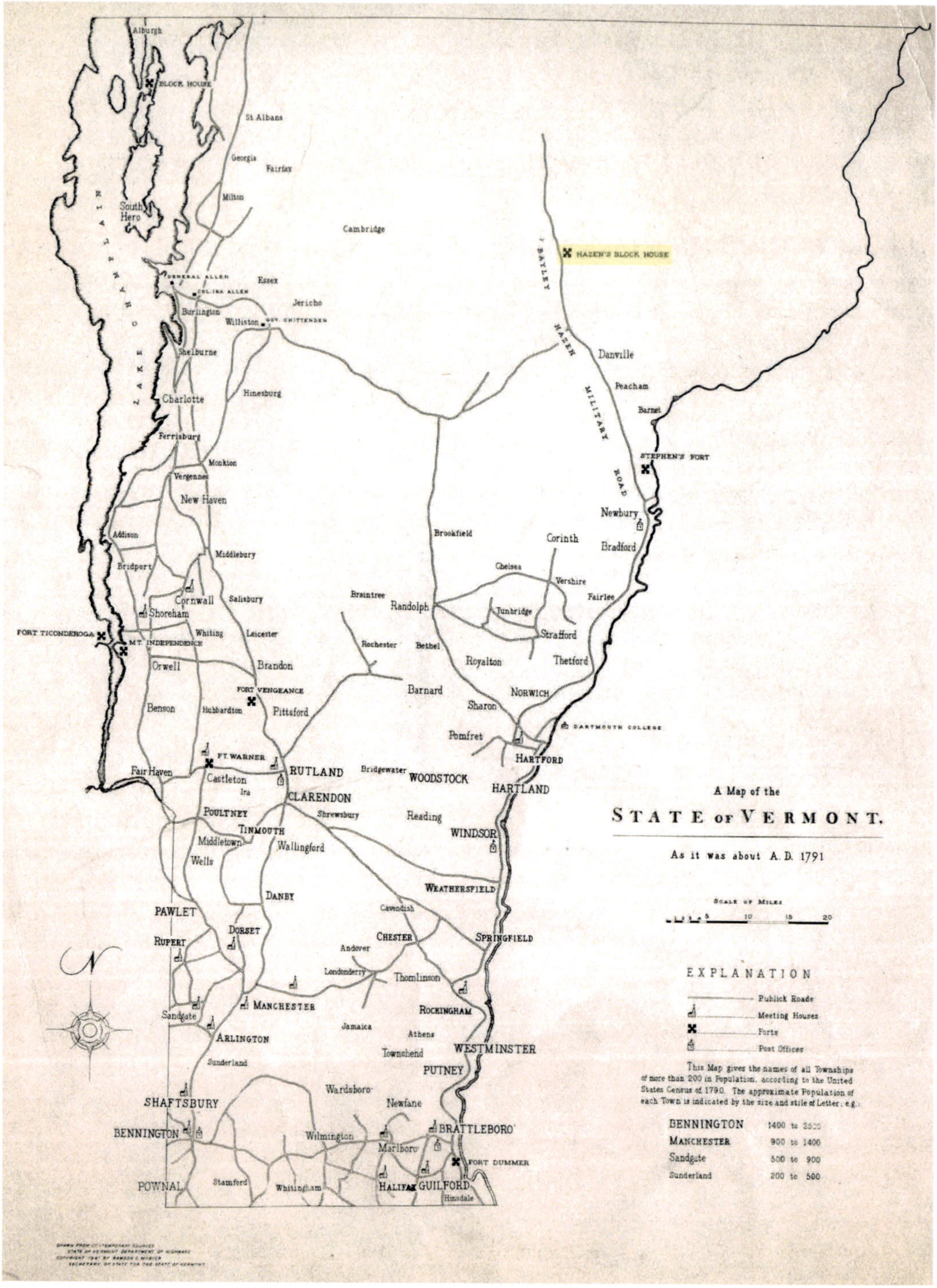

Figure 3.11. Wilgus map 1791. The Greensboro blockhouse identified as "Hazen's Block House" appears to be on the east side of the Bayley-Hazen Road. Adapted from State of Vermont Department of Highways.

To better understand the Greensboro blockhouse, it is important to set it into context with the other three. Although these blockhouses were separated by several miles, they operated in concert, and their architecture was similar and function-specific. In other words, their overall design and materials used were similar with some noticeable differences based on the personnel and activities they were intended to accommodate. Therefore, each of the four Bayley-Hazen Road blockhouses will be discussed individually below.

Peacham ~ First Blockhouse on the Bayley-Hazen Road (Figure 3.12)

- Approximate coordinates of the blockhouse: Latitude 44.304242, Longitude -72.158981
- The town of "Peachum" was originally a New Hampshire Grant chartered by Gov. Benning Wentworth in December 1763.
- "Peachum" was located in Orange County by February 2, 1781, and later in Caledonia County by 1792.

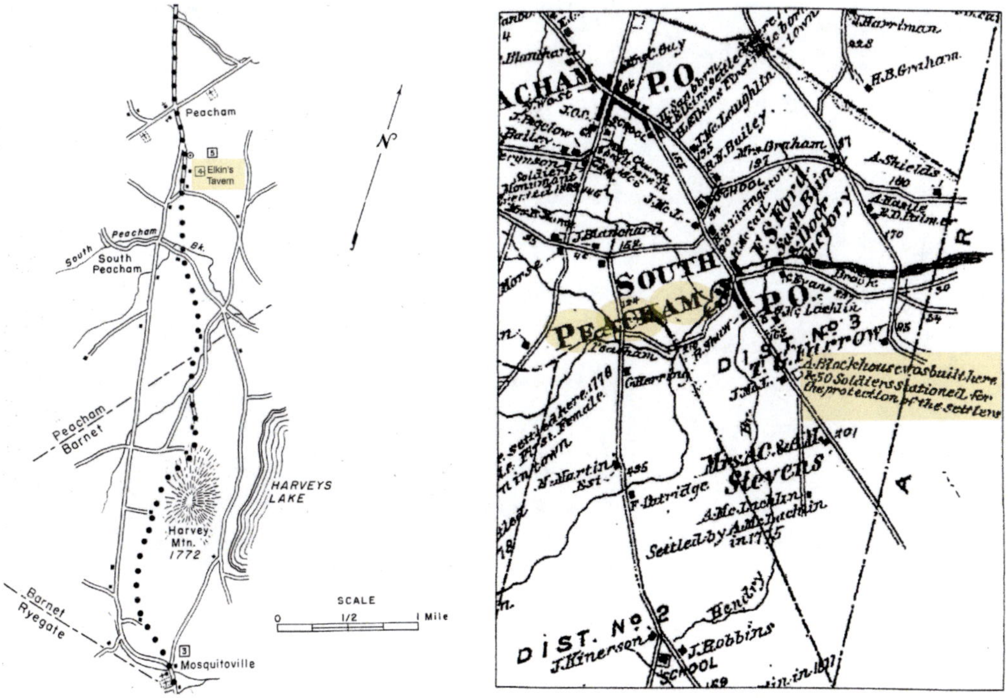

Figure 3.12. Early Maps of Peacham, Vermont. *Left*: English and McCorison, map of South Peacham with Elkin's Tavern highlighted in yellow. *Right*: Beer's map of South Peacham with the blockhouse site highlighted in yellow. Adapted from English and McCorison 1959:3 and Beers 1857, respectively.

The Peacham blockhouse, built in 1776, is the blockhouse about which we know the least. It was abandoned in mid-June 1776 when the first stage of the construction of the Bayley-Hazen Road was terminated by Gen. George Washington. Jonathan Elkins described this structure as being "a log house with port holes, which was called the first block house."[79] Unfortunately, little or no trace of this blockhouse remains, making it difficult to determine its original location. The only known cartographic reference to its location is the Beers *Atlas of Caledonia County, VT* (1875:15) (Figures 3.13). In the area labeled South Peacham, District 3, a square is drawn on the Bayley-Hazen Road with a notation that reads, "A Blockhouse was built here & 50 Soldiers stationed for the protection of the settlers." However, this notation was made well after its construction and ensuing events.

An inscribed roadside marker (Figure 3.14) erected by the Sons of the American Revolution in 1924 states, "The Peacham stockade built around the James Bailey (Bayley) House in 1780 stood in the field about ten rods to the east [of the marker]" (Figure 3.13). It also states, "The house of Jonathan Elkins (in 1776) stood on the opposite side of the road, about thirty-two rods to the south."[80]

Figure 3.13. *Left*: Close-up view of the Peacham map from Beers, *Atlas of Caledonia County, VT,* portion of the Peacham map, South Peacham. *Right*: View of the location of the stockade and the James Bagley cabin (1780). Photograph by J.P. Gill.

---

[79] Elkins 1921:192–193.
[80] Plumley 1924:2.

The present investigation of the Peacham stockade and blockhouse was aided greatly by the Peacham Historical Association who facilitated contact with Jock Gill of Peacham. Mr. Gill sent by email many photographs of the roadside marker for the blockhouse, the field where the James Bayley cabin and stockade site were located in 1780, the possible site of the blockhouse, and the Beers map (1875) locating the blockhouse in what is now South Peacham (Figures 3.13, 3.14).

This Tablet
Records The Following
Historical Sites

A Section of the
Hazen Military Road
Built by Gen. Jacob Bailey in 1776

The Peacham Stockade
Which Was Built
Around the Bailey House in 1780
Which Stood in the Field
About Ten Rods to the East

The House of Jonathan Elkins
Built in 1776, Which Stood
on the Opposite Side of the Road
About Thirty-Two Rods to the South

The Elkins Tavern
Built in 1787, Which Stands
About Thirty-Two Rods to the South

Erected by
The Vermont Society
Sons of the American Revolution
And the Peacham Historical Society
1924

Figure 3.14. Peacham roadside marker with inscription. Photograph courtesy of J. Gill.

We are indebted to him for his invaluable assistance. Jock and his wife Johanna Branson, the president of the Peacham Historical Association, live in the Jonathan Elkins tavern (1787), the oldest structure in town. Located on the Bayley-Hazen Road between Peacham and South Peacham, the Elkins Tavern was posted on the National Register of Historic Places in December 1978.

Mr. Gill comments: "My current view is that the Peacham blockhouse, built and abandoned in 1776, may have been in the field just north of the Tavern house where the 1924 marker says it was. This blockhouse would have protected Elkins cabin, one of the very few European houses in Peacham at that time. The stockade built in 1780 was reported by young Elkins as being 'a mile south of my father's.' I think I know where this is. Satellite imagery tends to support my hypothesis. It is probable that it was not in the horse pasture just to the east of Peacham-Groton road opposite what is now the Hartungs' horse farm, formerly the Harley Davis farm. At this site there appears to be evidence of a small cellar hole as well as water seeping out from what must be a spring" (Figure 3.13).

Mr. Gill further comments: "The 1780 fortification was built near where the majority of the seventeen or so settlers were located in 1780. This would be the southern parts of the 'square' and about a mile south of the Elkins log cabin, exactly as reported by Elkins. If you look at the names of settlers in 1780 and note their lot locations in the square as shown on the Elkins pocket map in the Peacham Town Hall, you can see how the above works. By 1780, the Elkins cabin would have been on the far northern edge of the settlement in Peacham." He concurs that there were two fortifications in Peacham, the early blockhouse soon abandoned, and four years later a picket around James Bayley's house a mile or so to the south of the abandoned blockhouse. "Consider also that the raiders who took young Elkins in 1781 would have had to march right past the old blockhouse site both coming and going," Gill adds, and he offers an interesting observation: "This would have been safe if it had been abandoned five years earlier. Their raid at the unfashionable late night hour may have been intended to avoid alarming the fifty troops stationed a mile to the south." The area where the original stockade or picket was likely erected around the log cabin of James Bayley (1779–1782) is now pasturage for horses and remains unexcavated, as is known thus far (Figure 3.14). The picket was said to have been built around Bayley's cabin in the spring of 1780 by a Capt. Aldrich.[81] So it could be concluded that two fortifications were built in Peacham during that period: the blockhouse in 1776 (though abandoned soon after when the road-building project was discontinued that same year) and the Bayley log cabin in 1780.

The Peacham blockhouse, although militarily abandoned, may have served as a stopover for travelers for several years, as did the blockhouses at Cabot, Walden, and Greensboro. Col. Frye Bayley, a nephew of Gen. Jacob Bayley, and Jonathan Elkins were two of the five pioneers credited with making the first pitches of lots in Peacham. As for the first settlers in Peacham, "Elkins has usually been credited with having been the pioneer settler, but this distinction should be shared with Frye Bailey."[82] F. Bailey writes, "… in 1773 I bought 500 acres of land in Peacham, labored on it three years, the first year I carried my provisions on my back there being nothing but a spotted line of trees between Newbury and Peacham. I built the first house in this town in 1776, and intended to have been the first settler."[83] He "agreed to undertake [a] journey"[84] to Montreal in February 1776 on orders from his uncle Jacob Bayley. Frye never resided in Peacham, but finally settled in Newbury after serving as a "scout, minute

[81] Baldwin, *The Vermonter*. 1906:309.
[82] Bogart 1948:40.
[83] Bailey 1926:30.
[84] Bogart 1948:40.

man, and guarding in sundry alarms."[85] The path that Frye Bailey's contingent took to Montreal in February 1776 would become the general direction of the Hazen Road.

Cabot Plain Blockhouse (Figure 3.15)

- Approximate coordinates of the blockhouse: Latitude 44.421938, Longitude -72.267483.
- Cabot was granted November 6, 1780; chartered by the Republic of Vermont to 66 proprietors August 17, 1781; surveyed and lotted by James Whitelaw in 1786. Cabot was at first part of Orange County and by 1814 part of Washington County.

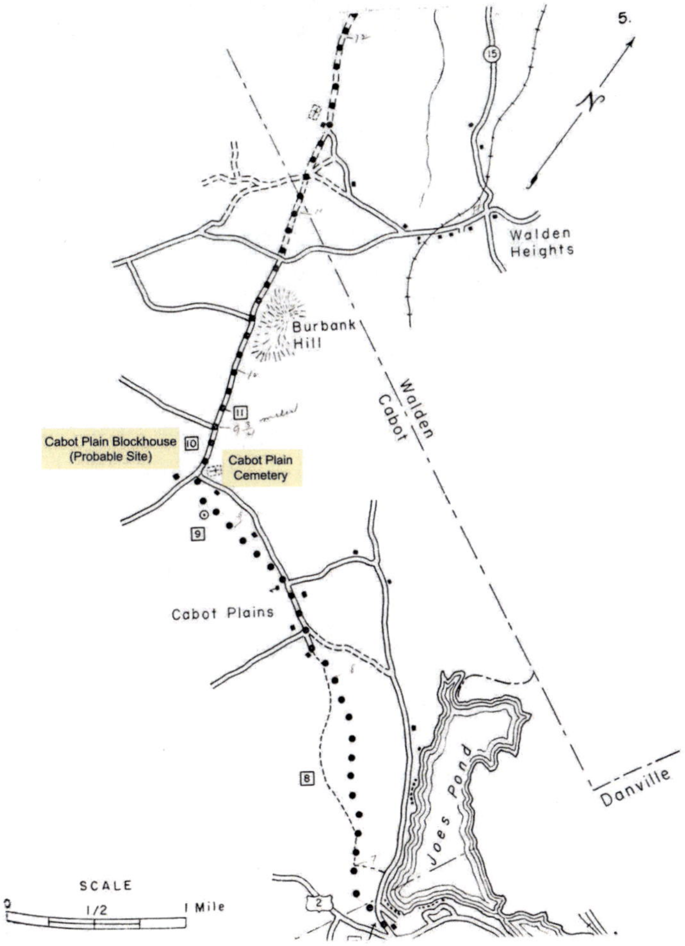

Figure 3.15. Map of Cabot, Vermont, locating the presumed site of the blockhouse and known cemetery site. Adapted from English and McCorison 1959:5.

[85] Bogart 1948:41.

The Bayley-Hazen Road extends through the northeastern corner of Cabot from the Peacham/Danville town line to the Walden town line, which is said to be about six miles from the Peacham blockhouse. The blockhouse on the Cabot Plain was on the highest elevation in town. It is worth noting that the Peacham blockhouse was mentioned in Elkins' *Reminiscences* and marked on the 1875 Beers *Atlas of Caledonia County, Vermont,* District #3 on the Peacham map, and clearly the first of the four blockhouses to be built, yet the Cabot blockhouse was described as the "first blockhouse."[86] The reason for this may be because it was the first one constructed after the restart of the road-building campaign in 1779. By that time, the Peacham blockhouse had been abandoned.

In 1779, General Hazen was ordered to Peacham with part of a regiment to complete the road commenced by General Bayley in 1776. Hazen had constructed a passable road 50 miles from Peacham through the northeast part of Cabot, over Cabot Plain, through Walden, Hardwick, Greensboro, Craftsbury, Albany, and Lowell to Westfield. Later, in March 1781, Col. Thomas Johnson of Newbury and Col. Jonathan Elkins of Peacham were taken captive by the British and marched to Canada. On their journey, "they camped the first night of this journey at the Cabot location, and when Col. Johnson returned on parole soon after, he again camped here, hence for many years after, it was known as Johnson's Plain."[87]

An early settler was Benjamin Webster (b. April 22, 1744, Kingston, New Hampshire), the uncle of the widely known orator Daniel Webster. Benjamin is said to have sheltered in the blockhouse while building his log cabin; however, no supporting textual references support this. Jane Brown, archivist for the Cabot Historical Society, offers her theory: "The site of Webster's homestead is about a mile from where we suppose the blockhouse was, so it would be unlikely he utilized that. Benjamin Webster's cabin was not far from Route 215, near the Cabot/Walden town line. However, Lt. Jonathan Heath's cabin was likely on or very near where the blockhouse was, across from the cemetery that is there today." Ms. Brown adds that Webster did not build separate buildings such as cabin, shed, barn, and outbuildings, but built one house that was "sufficiently capacious to answer for a house, barn, shed and all necessary outbuildings."[88] She continues by saying, "the early buildings were so crudely built, families would no doubt have frozen to death during winters had they not had the benefit of the body heat of each other and their animals all in one space."

On December 13, 1781, at Canterbury, New Hampshire, Benjamin Webster married Judith Heath, and in March 1783, he brought his family to Cabot.

---

[86] Elkins 1943:270–271.
[87] Child 1889:202.
[88] Hemenway 1882:76.

Benjamin drove the cow and Judith traveled on snowshoes, with a hired man pulling their few possessions on a hand sled (Figure 3.16). Two-year-old Hannah sat in a washtub on the sled. In the Hemenway *Gazetteer* account (1882:76–79), Hemenway claims that on their journey from Peacham to Cabot, the snow was four feet deep on the level areas. Benjamin Webster died October 19, 1827, at Walden, where he is buried in the Walden Heights Cemetery. Webster had served in Bedel's New Hampshire Regiment as a corporal.[89]

Figure 3.16. Hand sled pulled by Oliver Luce, first settler of Stowe, from Waterbury to Stowe, April 1794. Courtesy of the Stowe Historical Society.

According to Hemenway's *Vermont Historical Gazetteer* (1867:76–77), "the second settler to Cabot on the line of the Hazen Road was Lt. Jonathan Heath, who put down on the opposite of the present burying ground on the Plain" (Figure 3.17).

Figure 3.17. *Left*: Cabot Plain Cemetery. *Right Two*: The probable site of the blockhouse, across the road from the old cemetery. Photographs by Jill L. Baker, 2014.

---

89 White 1995:2917.

As stated previously, first settler Benjamin Webster (1744–1827) married Judith Heath, daughter of Caleb and Mary (Kezar) Heath, December 1781. Nathaniel Perkins Jr. (1754–1842), first settler to Walden, married Martha/Mercy Heath in 1775 at Canterbury, New Hampshire. It would be interesting to know whether these Heaths migrated together and how they were related to each other, a separate study in itself.

In an August 2012 correspondence, Jane Brown writes that to her knowledge, no surveys have been conducted in the area said to be that of the Cabot blockhouse (Figure 3.17). The Hemenway *Gazetteer* (1882:76–77) claims that the Cabot blockhouse was similar in size to the Walden blockhouse. There were stone markers on the military road for the location of the militia's winter camp and for the second location of the Yellow House Tavern (a.k.a. Smuggler's House),[90] but nothing about the blockhouse (Figure 3.18).

Figure 3.18. *Left*: Stele marking the encampment. *Right Two*: The stele marking Smuggler's House. Photographs by Mary L. Baker, 2014.

Brown writes, "That area has been plowed over and planted for many years so that any evidence of the first location for the tavern or the fort is likely lost." Jane grew up on the Plain, and the [Bayley-Hazen] Road crossed a good portion of their land east of where the blockhouse would have been. This area was called Fortification Hill, and the Plain was the hub of the town of Cabot for eighteen years.[91] From the encampment, only one artifact has been recovered: a large iron kettle that was likely used for cooking meals for the soldiers (Figure 3.19). We are grateful to Jane Brown for her input, photo, and label for this artifact, and for

[90] Hemenway 1882:78.
[91] Hemenway 1882:76–80.

providing sources of information about the Cabot blockhouse at the Plain.

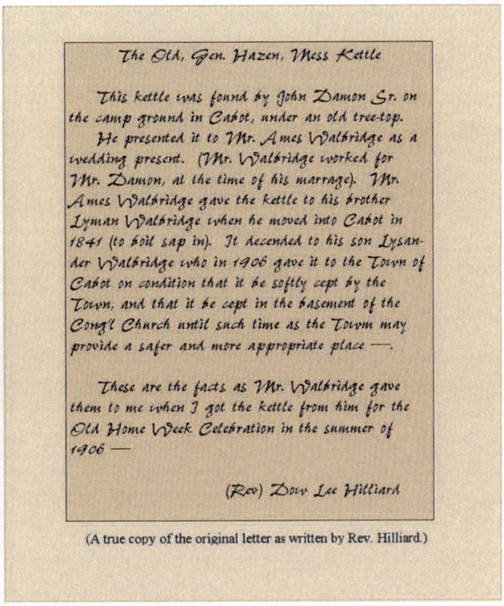

The Old, Gen. Hazen, Mess Kettle

This kettle was found by John Damon Sr. on the camp ground in Cabot, under an old tree-top.

He presented it to Mr. Ames Walbridge as a wedding present. (Mr. Walbridge worked for Mr. Damon, at the time of his marrage). Mr. Ames Walbridge gave the kettle to his brother Lyman Walbridge when he moved into Cabot in 1841 (to boil sap in). It decended to his son Lysander Walbridge who in 1906 gave it to the Town of Cabot on condition that it be softly cept by the Town, and that it be cept in the basement of the Cong'l Church until such time as the Towm may provide a safer and more appropriate place —.

These are the facts as Mr. Walbridge gave them to me when I got the kettle from him for the Old Home Week Celebration in the summer of 1906 —

(Rev) Dow Lee Hilliard

(A true copy of the original letter as written by Rev. Hilliard.)

Figure 3.19. *Left*: Cabot's large iron kettle used by Bayley's soldiers at the encampment with label/caption, now held at the Cabot Historical Society. *Right*: Description of kettle and its discovery by Rev. Hilliard. Photographs courtesy of Jane Brown, Nov. 15, 2015.

The Walden Blockhouse (Figure 3.20)

- Approximate coordinates of the blockhouse per Jess Robinson: Latitude 44.464163, Longitude -72.282356.
- Site # VT-CA-24 at Vermont Division for Historic Preservation, Montpelier, Vermont.
- No roadside marker. The site is in the woods on a gated logging trail off the Bayley-Hazen Road in Walden uphill about a mile from the South Walden Four Corners/Cabot Road transformer. Permission needed from landowner to visit the site. Lately used as a pasture or potato field.
- Walden was originally one of the lease land grants established by Gov. Benning Wentworth of New Hampshire on November 6, 1780; chartered August 18, 1781 by the Governor, Council, and General Assembly of Vermont in Orange County, and by 1792 in Caledonia County.

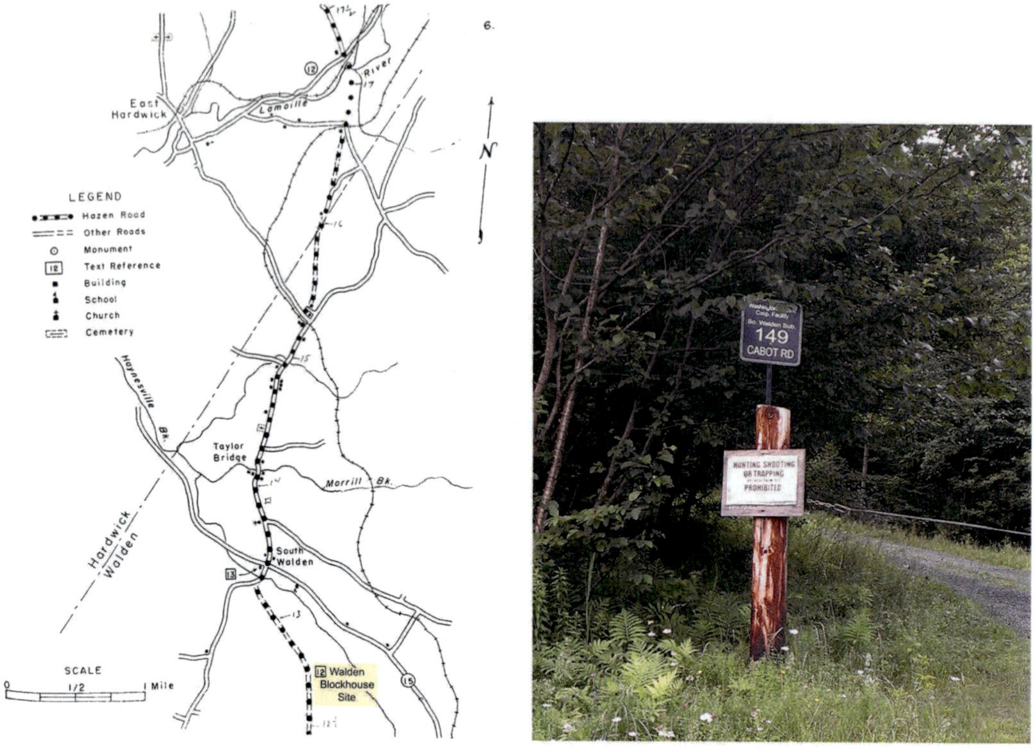

Figure 3.20. *Left*: Walden, Vermont portion of the Bayley-Hazen Military Road, with the blockhouse site marked. Adapted from English and McCorison 1959:6. *Right*: Logging trail leading to Walden blockhouse site. Photograph by Jill L. Baker, 2014.

Of the four blockhouses, the Walden blockhouse is the only one whose ruined foundations have survived above ground, thus allowing for evidential study and interpretation without archaeological excavation. Although some of the foundation stones have been removed throughout the years, those that remain *in situ* offer useful evidence for this blockhouse's layout and construction features. Based on the remaining foundation, the approximate size and shape of the blockhouse has been ascertained as well as the discovery of a stone-built water well. It is important to note that the water well was located partially inside a corner of the foundation walls offering protection to the water supply and those drawing from it. This suggests that each blockhouse may have included a similarly protected water supply, a well-known and ancient tactic. In 1985, P. Haslam and archaeologist M. Squire-Hackey surveyed the site, and a top plan was made of the existing foundation and well by Squire-Hackey (Figure 3.21).[92]

[92] See also Haslam 1997:9.

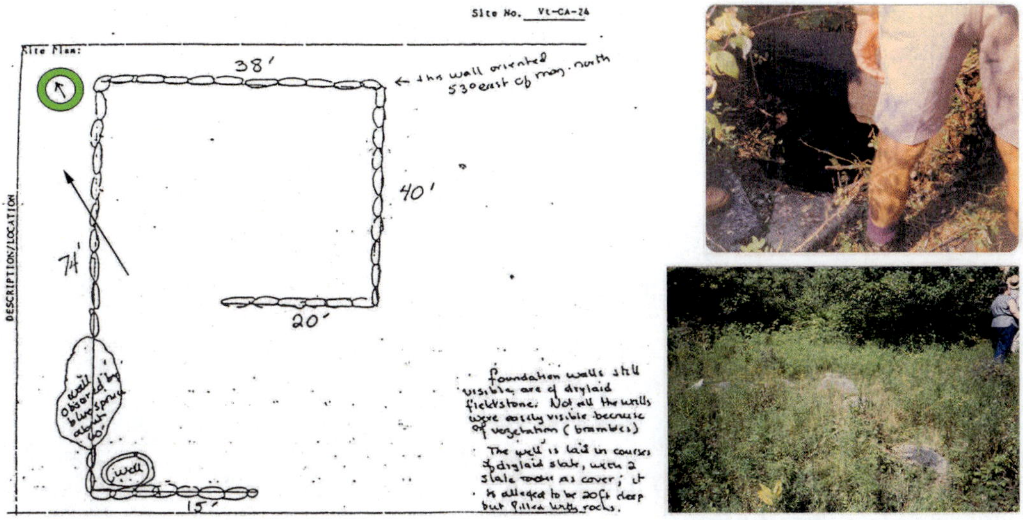

Figure 3.21. *Left*: Top plan of the remaining Walden blockhouse foundations and well. Drawing prepared by archaeologist Mariella Squire-Hackey, 1985. *Upper Right*: Water well located within the blockhouse. Well noted in the top plan of the Walden blockhouse, which P. L. Haslam also viewed. *Lower Right*: Remnant of foundations visible above ground. A large flat boulder approximately 4 feet long, 3 feet wide, and 1 foot thick lay in the middle of the space inside the walls. This large stone may have served as a hearthstone or as a food preparation surface, or as seating for more than one person. Photographs by Patricia Haslam, 1985.

Based on this survey, it is possible the well was dug first and the foundations established soon thereafter. In his paper, Maurice Eddy (1958–1959) suggests the well was dug and then walled-in by the garrison.

Blockhouses were usually built on a height of land to keep the site as dry as possible since there were numerous natural springs. To create the well, a wide hole was dug and stones were laid in dry, without mortar. Then, soil was backfilled against the stone, thus creating a well. The Walden well was protected on two sides by foundation stones, which served as footings for the walls of the blockhouse or for a stockade. This may suggest that the well was dug first and the foundation stones or walls positioned around it to protect it. Once the foundation and lower walls were built, the log superstructure was constructed. Assuming the other three blockhouses incorporated wells in the same manner, the extent of the blockhouses and stockade areas could be determined with more certainty.

Francis Foster, the landowner in 1976 and again in 1985, informed us that many of the blockhouse's foundation stones were removed and reused in the cellars of the first settlers' homes, specifically the Nathaniel Perkins and Smith families. The cellar hole of the Nathaniel Perkins house, one of the earliest settlers in Walden, is approximately 20 feet from the blockhouse, which the Perkins family used as their dwelling while building their log house. On August 23, 1976, during the nine-day bicentennial hike along the Bayley-Hazen Road from Newbury to Montgomery Center, Maurice Eddy, guide for the Walden portion, drew attention to a rose bush that still grows next to the trail marking where the militia buried one soldier. Near this grave the first cemetery in Walden was established, wherein the Perkins and Smith families were buried. Mr. Eddy also shared, "The entire garrison area was approximately a quarter of an acre. There was a second building inside the garrison, probably for a kitchen, dining, and armory. The well, 20 feet deep and 18 inches in diameter, is stone lined, laid up dry. There are stones filling in the well now, dropped in by children. Three flat stones are on top as a cover. There was probably a long rope and pail to haul water. The well was located near the apple tree and inside the garrison, but *not* inside the blockhouse. The blockhouse was used as a church and a school after the Revolutionary War."

Perhaps the best contemporaneous description of the Walden blockhouse was offered by Jonathan Elkins Jr. in his *Reminiscences* (1921): "[The blockhouse] was bult for to be a more servicible one, [than the one in Cabot] a large log house 20 by 40 feet, with a stone chimley in the midle, and the beams runn all around the house about 6 feet and covered with hewed timbers on the outer side of this House was carried up about 6 feet higher with port holes for small armes, and about 3 or 4 rods from the house thare was placed the tops and limbs of tres 8 or 10 feet high and sharpened to a point except a narrow way to pass in, and thare was brush sharpened in the same way to fill up the road if nessery, and all the trees was all fell for some distance from the blockhouse, Etc." If Elkins is correct in his measurements of 20 feet by 40 feet for the Walden blockhouse, the measurements taken by archaeologist Mariella Squire-Hakey of the site in 1985 (Figure 3.21) are of a larger structure: 38 feet wide and 74 feet long on one side to accommodate the water well. However, when measured from an opening to the inner portion of the structure without the well and surrounding wall, it measures approximately 38 feet by 40 feet. The water well extension of about 34 feet by 15 feet may have had a gate or a stockade across the opening originally. In any case, the Elkins and Squire-Hakey measurements are comparable.

In his work "The Bayley-Hazen Road" in *Vermont History* Vol. 27. No. 1, Marcus McCorison described the Greensboro blockhouse as "another blockhouse [that] was built at Caspian Lake similar to the simple Cabot one" (1959:60), which could imply that the Walden blockhouse was the largest and most substantial of the three blockhouses built in 1779.

Once decommissioned, the Walden blockhouse served as a civilian residence. The F. W. Beers *Atlas of Caledonia County, VT* map of Walden,

1875, notes in the southwest corner at South Walden that "N. Perkins settled here in 1780. Was the first settler in town." Mrs. Martha Perkins (sometimes called Mercy) gave birth to their third child, Jesse, on November 19, 1790, the first non-Native American born in Walden and born in the blockhouse (History Committee 1986:27, 37; *Vermont Vital Records*).

However, a contradictory entry in the *Vermont State Papers: Petitions* (Manuscript Vermont State Papers, Vermont State Archives and Vital Records Administration, MVSP, Vol.18:20, 165) states that on October 7, 1790, Jesse Leavenworth petitioned for an extension, granting more time in which to make settlement in Walden "when there were no settlers in town." Perhaps Leavenworth didn't know the Perkins family lived in the blockhouse, but that seems unlikely, or perhaps the Perkins family arrived between that date and the birth of their son Jesse on November 19, 1790. The 1790 Vermont Federal Census was taken in 1791, which states that only two families lived in Walden, Orange County at that time: Nathaniel Perkins with six in the family, and Samuel Samburn (Sanborn) with five family members.

Jesse Leavenworth is listed in nearby Danville, Orange County. Nathaniel Perkins Jr. was born 1754 at Canterbury, New Hampshire, and served from New Hampshire in Col. Timothy Bedel's Regiment, Capt. Hutchins' Company, and was at the Battle of Bunker Hill serving with Col. John Stark during the Revolution. Perkins married Martha/Mercy Heath in 1785 at Canterbury, New Hampshire. After settling in Walden in 1789, he was the first town clerk and one of the first Selectmen. The first sermon was delivered at his house, and he taught the first school in the blockhouse in 1796. In 1795 he was elected a town representative to the Legislature. He and his wife Martha are buried in the cemetery at Walden Heights.

Greensboro Blockhouse (Figure 3.22)

- Approximate coordinates of the blockhouse: Latitude 44 34' 38.56" N, Longitude 72 18'59.50" W
- Greensboro was chartered August 20, 1781, by the Governor, Council, and General Assembly of State of Vermont.
- Site number VT-OL-71, registered at Vermont Division for Historic Preservation, Montpelier, Vermont.

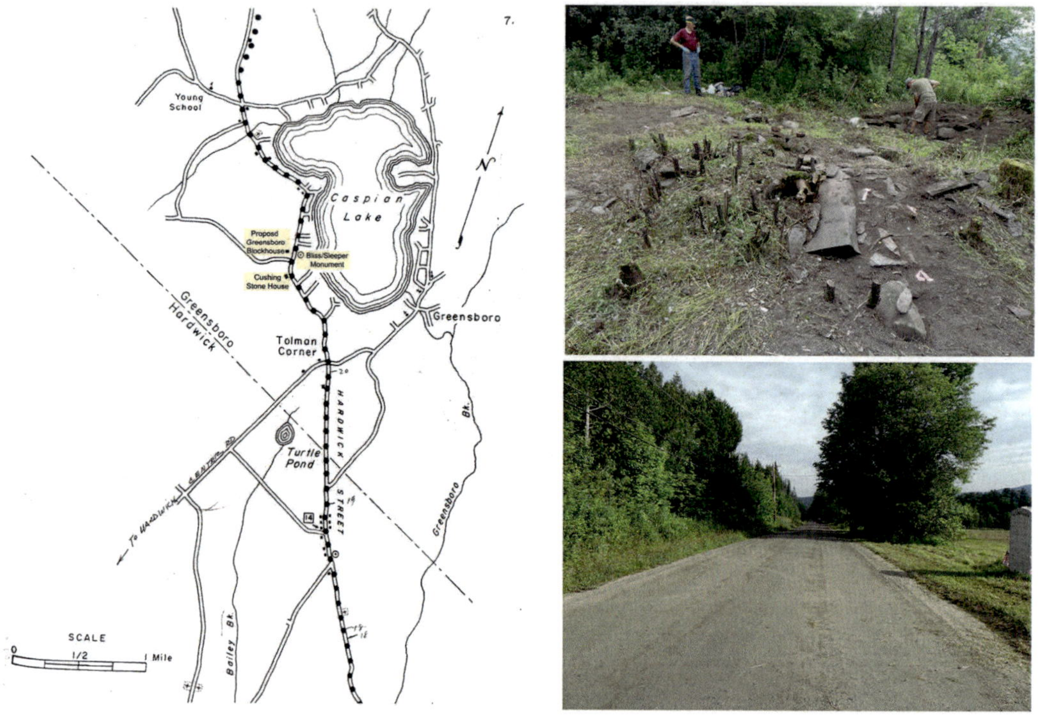

Figure 3.22. *Left*: Map of Greensboro with the proposed blockhouse site, the Bliss and Sleeper monument, and the J. Cushing house marked. Adapted from English and McCorison 1959:7. *Right Upper*: Greensboro blockhouse site. *Right Lower*: Bayley-Hazen Road near Bliss and Sleeper monument. Photographs by Jill L. Baker, 2010.

The Greensboro blockhouse was the fourth and last defensive structure erected to guard the Bayley-Hazen Road. It is important to remember that on the 1790 and 1794 maps by James Whitelaw, the blockhouse is located on the *east* side of the Bayley-Hazen Road, and on the Wilgus map of 1791 and the Bohn map of 1796, the blockhouse at Greensboro is on the *west* side of the Bayley-Hazen Road. This discrepancy may be due to a re-routing of the road sometime between 1794 and 1796, which was noted in the Administrator's deed of Thomas Durkee, on behalf of the Estate of Aaron Shepard, deceased, in papers filed at the Orleans County Probate Court documenting the sale of the Shepard farm to Joel Cushing on October 21, 1813.[93] The 110-acre property is described therein, "excepting about five acres which was deeded by the said Aaron Shepard to Enoch Tolman 29 April 1811 the western line of which piece is . . . the line of the new road or the road as it now runs instead of the Old road extending the piece westerly to the new road . . . ." Aaron Shepard died in May 1811.[94] The location of the blockhouse on the Bohn map seems to be at odds with the location of the road today, whereby the road is on the east side of the blockhouse/homestead.

As a result of our preliminary investigation (2010–2015), the proposed site of the blockhouse, which was, by spring of 1789, the Aaron Shepard homestead with evidence of homesteading, testifies to the fact that the road is on the *east* side of the blockhouse/homestead, and thus the old road would have been to the west of the blockhouse. Unfortunately, this homestead does not appear on any map. Indeed, a lack of vegetation on a wide path just a few feet from the northwest corner of the blockhouse site, and continuing north for several yards, could indicate a former roadbed. It could also have been a farmyard for animals. This includes an old well to the north, now covered, but has a flattened barbed wire and rotting wood fence surrounding it. The well was found to be stone-lined before the new cover was put on by the property owner in 2013.[95] The Shepard farm may have expanded over the early years to include outbuildings and other needs for subsistence farming and could have encompassed this well. In 1984 Raymond Mercier, a previous owner of the property on which the blockhouse/homestead is thought to be situated, drew a rough penciled circle on a modern map and labeled it "old well" (Caspian Greenland Corp. Prepared by R. N. Bohlen 1986, see Figure 2.2, Chapter 2) at the corner of the Lakeshore Road (formerly the Bayley-Hazen Road) and Cook Hill Road. At that time, Mercier said that the well was stone-lined; he was not sure how deep, "but cattle got into it in the pasture so it had to be filled in. There is nothing to be seen now." This well has not been rediscovered at the time of this writing, but must have been within the immediate blockhouse and stockade surround.

In 1959 Dan and Thelma Haslam purchased the property known today as 197 Blockhouse Hill Road in Greensboro, Vermont. It was first owned by Dr. David Marvin of Essex Junction, Vermont, a physician, and later a member of the medical faculty at the University of Vermont and a bank president in 1940, who built a cottage and garage/icehouse in 1923. After purchasing the property, Dan and Thelma found a tin wood-framed sign

[93] Greensboro Land Records A:451.

[94] Greensboro Congregational Church Death Records Vol. 1:178; Stone 1854.

[95] It should be noted that the State of Vermont, Agency of Natural Resources, shows two wells for this area in their database: one located just southwest of the proposed blockhouse site labeled as Well Report Number 42 completed in 1974 and another well located just north of the site labeled as Well Report Number 32402 completed in 2007. It is not known whether these two wells correspond in any way with the two wells discussed here.

hanging in the two-car garage/icehouse, which reads Block House Hill (Figure 3.23). It is not known when this sign was made or where it originally was posted, but it is pockmarked with bullet holes and has weathered; it is now displayed in the Haslam cottage. Doctor Marvin was no doubt responsible for posting the sign each summer at the juncture where Block House Hill meets Lakeshore Road as a way to direct travelers to this popular vacation spot, which was first called the "Barre Camps" in the early 1920s when B. Hooker of Peacham and Barre built a cottage there along with his son-in-law H. Cutler (Figure 3.23).

Figure 3.23. Large "Block House Hill" double-sided hanging sign, tin with wood frame. This sign was found in 1959 in the garage/icehouse of the David Marvin camp property, now 197 Block House Hill Road, when Daniel B. and Thelma Haslam purchased the camp. Photographs by Jill L. Baker, 2016.

The hillside location for the Greensboro blockhouse, overlooking Caspian Lake, was considerably rockier than the other three blockhouse sites. It is likely that the first floor walls were constructed of stone simply because of the availability of stone as a building material. Furthermore, using the available stone cleared the land for habitation. Boulders and rocks were scattered here millennia ago by receding glaciers. In fact, the nearby Stone House, situated 250 yards to the south on the Bayley-Hazen Road (now the Lakeshore Road), "was built from one boulder on the site with material left over" before 1816.[96]

It was said by unnamed past researchers and archaeologists that 'corduroy-laid logs' were found in short sections of the road, but their exact locations remain unknown. It would be useful to know if any sections of corduroy road exist near the Greensboro site. If so, these may help to interpret the original location of the road and the discrepancy between the maps that locate the Greensboro blockhouse on different sides of the Bayley-Hazen Road.

Across the road from the proposed blockhouse site, on the east side of the road, stands a large granite roadside marker with the inscription "1781–

[96] Thompson 1848:75; Haslam 1996:16.

Near This Spot By A Block House Guarding Hazen Road Two Scouts Constant Bliss and Moses Sleeper Were Killed By Indians And Buried Where They Fell. Lest We Forget The Pioneers This Memorial Was Erected 1941" (Figure 3.24). This commemorative marker is viewed nearly every summer day by walkers, cyclists, and others traveling the road. It stands on a five-foot-deep foundation located on a 32-foot x 60-foot parcel of land owned by the town, which maintains the lawn around the monument during the summer and places two small American flags there.

Figure 3.24. Greensboro, Vermont, monument to the memory of Constant Bliss and Moses Sleeper. Erected 1941. Photograph by Jill L. Baker, 2010.

Summary

Originally the four blockhouses would have functioned in concert to protect the military road. The Peacham blockhouse is the site about which we know the least and was the least used of the four blockhouses; however, J. Gill has provided new data that will provide some direction to future investigators. The approximate location of the blockhouse at Cabot Plain has been identified by J. Brown, archivist for the Cabot Historical Society. Together with the encampment site, future archaeological excavation could unveil historic activities there as well.

More is known about the Walden site and its exact location; the foundation that protrudes aboveground provides a footprint for the size and shape of the structure and the location of the water well. However, only through archaeological investigations will its habitation activity come to light. If this site were ever investigated archaeologically and scientifically, the nearby Perkins-Smith settlement, literally a stone's throw away, should be included in such a study as well.

The Walden blockhouse may have been the most substantial of the four blockhouses, and could provide the most complete corpus of data when studied in conjunction with the contemporaneous settlements nearby. Perhaps archaeological investigation will be possible here in the future. The site proposed for the Greensboro blockhouse also shows promise. The rock pile that was probed from 2009 to 2015 has yielded important data, which will be discussed below. For all the blockhouse sites, future investigations could make use of drones and ground-penetrating radar or sonar that may provide useful data for all the blockhouses.

It is important to note that while the initial purpose of blockhouse construction was military, several blockhouses continued to be useful after the war. Upon decommission, the early settlers, mostly from New Hampshire, utilized them as stepping-stones on their way to the towns in which they finally settled. The structures were used as homesteads, churches and schools, and probably for town business meetings. They formed the nuclei out of which grew the early neighborhoods. Additionally, the Bayley-Hazen Road provided a conduit not only for early settlers but also for the transportation of raw and finished goods and services. While this road may never have been completed nor fully served its intended purpose, its long-lasting overall function and value remain evident.

Chapter 4

The Greensboro Blockhouse Site: Archaeological Method and Stratigraphic Analysis

Excavating an ancient or historic site involves more than simply digging holes in the ground to find cool stuff. Archaeology is not a treasure hunt; rather, it is the controlled investigation of stratigraphy, artifacts, and architecture so that once removed from its original context, the dismantled components can be reassembled for analytical and interpretive purposes. The location and context of artifacts and architecture provide details regarding the use of objects, the construction of structures, and human activity, and they offer clues about the people who inhabited the site. To better understand the Greensboro blockhouse site, the initial task was to establish a methodological approach to the exploration of it, after which excavation and analysis could begin. This chapter will present the methodological approach adopted by Archaeological Horizons, Inc. (AHI) and a stratigraphic analysis.

Our methodology was determined by a combination of factors: research goals, archaeological ethics and method, and limitations outlined by the property owners, the last of which heavily influenced the direction of the project. These constraints dictated that the exact location of the site could not be disclosed in an effort to maintain the privacy of the family members who reside on the property and to protect the site from would-be looters.

It was also stipulated that no portion of the site could be restored. While the property owners were happy to conduct full-scale excavation at the site, they were not in favor of any type of restoration, especially if that would attract visitors and/or devalue the property itself.

Since archaeological excavation is inherently destructive, post-excavation preservation is essential to provide an educational explanation to the public regarding the function and significance of a site. Given these constraints, AHI devised a way to maintain professional, ethical integrity while honoring the wishes of the property owners. Overall, it was agreed that determining the true nature and history of this ruin was important, as was regaining knowledge that had been known to earlier generations but lost to recent ones. If this was, in fact, the blockhouse, it could be protected from possible future destruction. The only way to do this, of course, was to excavate. The research goals and excavation plan included a strategy of compromise that would help to gather important data while remaining minimally invasive, and thereby maintain Mother Nature's excellent preservation of the site.

The research goals for the project were three-fold: determine the original purpose of the building (was the collapsed ruin an undocumented farmhouse or the blockhouse?), identify the military and civilian phases, and determine the extent of the structure. To do this it was necessary to understand the structure's architectural features and interior configuration, and to analyze any artifacts (material culture and ceramics) that may have been left behind by the inhabitants. These goals were achieved through careful archaeological excavation of several crucial areas of the site. The data collected from excavation helped to determine its original purpose as a military blockhouse and its secondary use as a residential dwelling.

To collect data while preserving the overall integrity of the site, a surgical approach was adopted by opening a series of probes that provided a glimpse into the character of the structure, and those who inhabited it, without inflicting too much damage to the site. An overall surface survey was conducted in 2010, and two probes and one square[97] were excavated in 2011, 2012, and 2015 respectively. During these focused investigations, a grid was established, extensive field notes were recorded, architectural features and loci were fixed onto top plans, photographs were taken, and artifacts were collected, photographed, and cataloged. At the end of every season landscaper's cloth was laid across the excavated area, and all the rocks and soil that were excavated were backfilled into each probe and square. In this way, future excavators will know exactly where AHI conducted its investigation of the site. Backfilling also helps to maintain the integrity of the site. To abide by the property owners' wishes, the exact location of the site will not be disclosed in this or any publication. Only the general location will be mentioned as being in the

[97] Prior to excavation, a grid is established across the site in increments of 5 x 5-meter squares. A probe is the partial excavation of a square.

vicinity of the monument memorializing Constant Bliss and Moses Sleeper on Lake Shore Road in Greensboro, Vermont.

The surface survey conducted in July 2010 revealed the general parameters of the building. At least three of the walls — north, east, and west — were easily identified from the surface without excavation. Visible within the walls were numerous rocks that likely represented much of the collapsed building. Artifacts collected from the surface included bricks, nails, a hoe head, and a metal disc (probably the base of a vessel). Additionally, several enticing architectural features were discovered. Based on these results, it was determined that the rubble represented a collapsed building that was once inhabited and was worthy of excavation. It was extremely tempting to excavate the areas where the most interesting architectural elements were evident. However, the areas chosen for excavation would yield valuable information yet reserve the excavation of key elements for a future time when a preservation plan could be established. Furthermore, archaeological method also dictates that one does not excavate the whole of a site, but leaves a portion of it to future generations who may have better technology, additional knowledge of the chronological period under investigation, and, of course, much better funding. With this approach we were able to obtain valuable information about the site while preserving its integrity.

Prior to excavating, the first task was to establish an artificial grid across the site (Appendix 2) for documenting excavation activity. In July 2011, with the help of a professional surveyor, Tim Short, and his wife Deborah Holmes, a 15 x 15-meter grid was staked out across the area presumed to be the collapsed structure. A benchmark (starting point) was established just beyond the north wall identified in the 2010 surface survey. From there the grid lines were established relative to magnetic north and extended 15 meters to the south and west with stakes at every 5 meters. (All measurements were metric.) It is within this grid that archaeological excavation took place: in Squares NE7 in 2011, NE3 in 2012 and NE4 in 2015 (Appendix 2).

Stratigraphic Analysis

The overall goal of the project was to determine whether the structure was in fact the blockhouse or an undocumented farmhouse and to identify and understand the military and residential phases of its habitation. Factors that determined the original purpose included architectural features, the layout of the structure, and the type of artifacts. Although this section may seem technical, detailed, and downright tedious, it will showcase the raw data that helped to formulate our conclusions. The following is a discussion of the stratigraphy (the layers and features that comprise the site) and associated artifacts which, when combined, establish a preliminary scenario for the structure.

The Second Preliminary Excavation Season[98], 2011

The first probe (Probe 1) was excavated in 2011. In addition to the project's overall goals, the goal of the 2011 season was to investigate the southern extent of the structure. The probe was situated in Square NE7, ultimately between the 10 and 13-meter marks[99] on the western north-south grid line, and extended 2 meters eastward into the center of the grid (Appendix 2). No sooner had we sunk a pick into the ground then the first nail appeared (Reg. 25)[100], though it was fragmentary. Soon after, only a few centimeters below topsoil, multiple rocks appeared. These were located in the southern half of the probe and extended across it diagonally. The rock feature (Locus 1) enters the probe at approximately the 10- to 11.5-meter mark along the western gridline and extends diagonally to the southeast corner (Figure 4.25).

Once articulated, it was clear that the rock feature consisted of unhewn fieldstones, mostly shale with one very large stone boasting a mostly flat surface on the exposed side. This feature formed a relatively straight line running in a southeast-northwest direction at a roughly 70-degree angle to the north-south grid line. Locus 1 was no more than two courses and measured approximately 0.22 cm high, above the surface of Locus 3. No mortar was found to hold the stones together; however, some hard-packed soil appeared to hold them in place. Multiple smaller stones, leveling stones among the larger ones, seem to have provided stability and made the top of the feature level. Associated with the rock feature, at the 10.8-meter mark along the western gridline was a pair of scissors (Reg. 45) located where the rock feature meets the western balk.[101]

---

98 Other than the surface survey (2010), the excavation seasons are described as preliminary. This is to differentiate between full-scale investigation and probing. Although excavation took place, the entire site has not been fully explored, but research has begun. The surface survey is considered to be the first preliminary season.

99 The probe was originally planned to be a 2 x 2-meter square, but was extended to 3 x 2 meters.

100 Our original intention was to include a Catalog of Artifacts, wherein readers could view all artifacts according to their registration number. However, present funding does not allow us to include a Catalog of Artifacts in this publication. Should funding become available in the future, AHI will publish a supplementary volume. It is our intention to share the artefactual results of our investigation, so for those who are interested please contact AHI at archaeologicalhorizons@gmail.com.

101 In archaeological terms, balk(s) refers to the unexcavated side wall(s) of an excavated square or the unexcavated portion of earth left as a partition between squares. The material left within balks helps with stratigraphic interpretation.

Figure 4.25. The 2 x 2-meter probe. *Left*: view to the northeast showing Locus 1, the rock feature, and Locus 2. *Middle*: view to the south showing Loci 1 and 2. *Right*: view to the south showing Loci 1, 3 and 6. Photographs by Jill L. Baker.

North of the rock feature was Locus 2[102], a fill layer comprised of black soil with inclusions of small and medium sized stones which appeared to have been floating and did not form any observable pattern (Figure 4.25). Multiple nails and ceramic fragments were also found within the matrix of Locus 2. The majority of nails were located north of the rock feature in the northwestern quadrant of the probe. Some nails were also located on or close to the rock feature. Curiously, most of the nails were found in pairs, one next to the other, consisting of a long and short nail. The ceramic fragments, forming teacups, a shallow bowl, and a plate (Regs. 29, 30, 51–55, 57, 96–99) as well as several terra cotta fragments (Regs. 56/95, 61, 89, 92, 94) were found in proximity to the rock feature on its northern and southern sides. Brick fragments were also scattered throughout the layer. Finally, multiple shards of glass (Reg. 31) were discovered in proximity to the rock feature, mostly to the south and southwest of it. These shards were greenish in color and flat.

Several other fill layers were situated around Loci 1 and 2. Immediately below Locus 2 was Locus 3 (Figure 4.25), a fill layer composed of light brown clay-like material. It was hard-packed with inclusions of small stones and clumps of darker clay. This fill layer also extended under Locus 1, the rock feature. Found within this layer were several nails, ceramic sherds, and glass shards, though not in the same quantity as in Locus 2. One whole brick was discovered in this layer next to the eastern balk in proximity to Locus 1, just north of it. On one of the broad faces of the brick was some burnt residue, and near the brick, a few small pieces of burnt wood and charcoal bits. These were collected and saved for future C14 analysis. Locus 3 was approximately 5 to 10 cm deep. Another hard-packed fill layer of gray soil, Locus 6, was situated directly below Locus 3. In addition to this layer, there were inclusions of shale as well as small and medium-sized stones. No pottery or material culture was found in this layer.

---

[102] Locus 2, on the northern side of Locus 1 (rock feature) is the same as Locus 4, located south of Locus 1.

In the southwestern corner of the probe, south of the rock feature (Locus 1) was another fill layer, Locus 4. The soil material that comprised Locus 4 was the same as that of Locus 2, and probably represented the same filling layer, but was separated by the rock feature. To identify specific areas within the excavation area, each location received a different locus number, even though these were the same filling layer. During post excavation analysis, these layers were combined. The same is true for Loci 5 and 3. A thick layer of gray hard-packed clay with inclusions of small pieces of shale, stones, and fist-sized stones was discovered directly below Locus 6. No pottery or material culture was found in this layer, and it probably represents virgin soil. With the discovery of this layer and its lack of evidence of any human activity, further vertical excavation was abandoned.

With time remaining in the dig season, we decided to expand the probe horizontally by one meter to the south, thus extending the probe to 2 meters wide east to west and 3 meters long north to south. The loci referred to above also extended into this probe. Initially a 2 x 2-meter probe was opened to better understand the southern extent of the site. However, because that area was so small and all the loci extended into the balk, we decided to extend the probe by one more meter to the south to better understand the loci and their stratigraphic relationships. The expanded probe ultimately measured 3 meters north to south and 2 meters east to west. As we peeled away topsoil, we discovered rocks in a seemingly random floating pattern, as if they had once formed a floor joist, a partition wall, or even a collapsed exterior wall. However, once articulated, it became clear that this was a one- to two-course feature and was relatively straight. Out of curiosity, we placed a board and level on top of the stones; remarkably, the stone feature was almost level (Figure 4.26).

Figure 4.26. Board and level placed on top of the stone feature (left) revealing that the stone feature was almost level (right). Photographs by Jill L. Baker.

After removing Locus 2 and articulating the rock feature (Locus 1), part of the stone feature was removed to reveal that it rested on Locus 3, which in turn rested on Locus 6. This facilitated a better understanding of the relationship between Loci 1 and 3. Locus 3 lifted off of Locus 6 with no difficulty, and it was a very different color and consistency from that of Loci 6 and 2. Based on the stratigraphic relationship, it could be concluded that Locus 3 served as a leveling layer for Locus 1. Upon expanding the probe, a large stone appeared, which measured approximately 1.2 meters wide x 1.4 meters long. This stone was made of shale, and the exposed side was relatively flat. (Figure 4.27)

Figure 4.27. Multiple views of Locus 1 after removing half of it to better understand the underlying stratigraphy in the expanded probe. The upper left is a view to the south looking at the northern side of the rock feature. The upper right shows the detail of the largest shale stone in Locus 1. The lower left is a view to the southeast and the lower right is a view to the west showing Locus 1. Photographs by Jill L. Baker.

The rock feature (Locus 1) likely represents a structural component to the building. In a discussion about early American building techniques, Roberts illustrates the foundation of the Thomas Clarence house in Manton, Rhode Island, ca. 1680, which boasts a stone foundation and a floor joist made of fieldstone with a wooden beam sitting on top of it. This joist spans the width of the main room and supports a timber frame for the flooring.[103] Locus 1 extended across the probe in a southeast-northwesterly direction into both the eastern and western balks, and was likely a floor joist; however, it posed several questions. Did the angle of the joist suggest the overall orientation of the building? Was this a sill/threshold representing an exterior wall/entrance into the building or an interior partition wall? Was it a floor joist? Did Locus 1 span the entire diagonal width of the structure or stop partway across? By speculating a possible trajectory for Locus 1 and assuming it is uninterrupted, the rock feature would eventually join the presumed eastern wall of the structure (Figure 4.28).

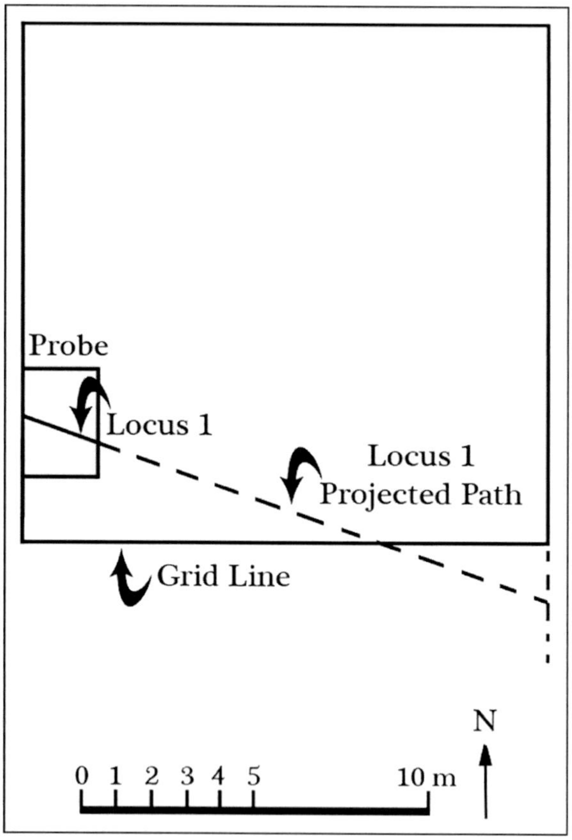

Figure 4.28. The existing grid and probe lines indicate the approximate location of Locus 1 within the probe. Locus 1 extended into both the eastern and western balks of the probe. This plan speculates the projected trajectory for the rock feature if it continued uninterrupted to the approximate location of the presumed eastern wall of the structure. Top plan by Jill L. Baker.

[103] The American Cities and Technology Reader: Wilderness to Wired City. London/NY: Routledge. 1999:64 Fig. 9.2.

Locus 1, the rock feature, likely represents either the southern outer wall or a floor joist within the interior of the structure.[104] Given the width of this feature, it could be considered an outer wall; however, its depth does not match that of the wall discovered in Probe 2 in 2012, which is definitely an exterior foundation/outer wall.

Therefore, it is most likely that this feature functioned as a floor joist that supported wooden flooring within the interior of the structure. Thus, when considered together, the cleared and leveled earthen surface (Locus 6), the light brown packed clay-like layer (Locus 3), and the stone feature (Locus 1) may have been sub-floor structuring. The nails found in the loose soil fill (Locus 2) may attest to the wooden beams and floorboards that once rested atop the structuring.

During the course of excavating the probe, 91 objects were recovered; 88 of these were registered and 3 were not registered due to their find spots. Of these, 53 (60%) were nails (Regs. 26–28, 32–44, 46–50, 59–60, 63–88, 100–105, 109–112). Of the 53 nails, 34 (64%) were bent. They ranged in size from 2.5 to 8.5 cm long and seemed to be found in pairs: one long and one short. The other items included ceramic fragments (some joining) (Regs. 29–30, 51–57, 89, 92–99), glass shards, animal bone (possibly worked) (Reg. 58), a pair of scissors (Reg. 45), burnt wood (stick not a board), and a brick (Reg. 108) with burnt bits stuck to one side. The majority of these items were discovered in association with the rock feature (Locus 1) both on and immediately north of it and in Locus 2. Few objects were recovered from Locus 3, and those that were lay generally close to the top of this layer.

The Third Preliminary Season, 2012

The goals of the 2012 season (Probe 2) were to determine the northern extent of the structure and to examine the foundation/first floor wall to better understand the primary purpose of the structure. Based on data collected from the 2010 surface survey, it was determined that the most advantageous location for examining the foundation would be in Square NE3 (Appendix 2). To maintain the structural integrity of the building, the probe was established slightly west of what is considered the northeast corner.

In 2010, the presumed northeastern corner of the structure was identified, and from this point the eastern wall, extending in a southerly direction, and the north wall, stretching in a westerly direction, were easily detected. Given the vulnerability of the eastern wall due to the adjacent drainage ditch that runs the length of Lake Shore Road, the northern wall offered the best option for the location of the 2012 probe. By

[104] I am grateful to B. Gauthier and his two colleagues who visited the site during excavation, and our discussion regarding the possibility of this feature functioning as a floor joist.

investigating the structural design, including the architectural layout as well as the size and width of the foundation and construction materials, it is possible to determine a building's character. According to the Dictionary of Fortification, a blockhouse would be constructed of materials that were "capable of sustaining and absorbing multiple bullet impacts on their walls and roofs" and that would "maintain their structural integrity under stress of severe lateral shocks."[105] Such a solid structure would require a foundation of commensurate density characterized by a substantial width and the use of unyielding construction materials. Clearly, the foundation of a blockhouse designed to support a solid superstructure and to withstand siege would be considerably more substantial than that of a farmhouse or barn. Consequently, our goal was to expose, examine, and identify a portion of an exterior wall and foundation.

Using the grid established in the 2011 season, the probe was located between the 3rd and 4th meter marks along the northernmost grid line in square NE3. From the 3rd and 4th meter marks on the northern grid line, the probe extended southward initially to the 3rd-meter mark and was subsequently extended to the 3.5-meter mark (Appendix 2). The entire probe measured 1 meter east to west and 3.5 meters north to south. Once the immediate area had been prepared and the parameters of the probe established, excavation of the area to the north of the stones visible on the surface was commenced to expose the northernmost extent of the wall and its exterior facing. From there excavation extended southward toward the interior of the building, exposing the width of the wall and some of the interior floor space. To reiterate, the purpose of this probe was to determine the width and composition of the exterior wall.

After clearing away the vegetation, the topsoil was removed to access historic layers. Depending on the location within the probe, the topsoil was approximately 3 to 5 cm deep. There was a slight upward curvature in the topsoil toward the stone wall before tapering off in a northerly direction. It was observed that the topsoil was deeper on the upward curve and shallower toward the southern side, closer to the visible stones. As the topsoil approached the exposed rocks, it tapered off and filtered into the spaces between the rocks. The elevation of the topsoil was approximately 455 meters above sea level according to Garmin's Navi GPS. Furthermore, about 0.50 cm into the probe south of the grid line, the elevation of the topsoil was higher than the surrounding forest floor, which appeared to be 1 to 1.5 meters lower. It is unfortunate that these elevations are only estimates, but the limited resources of AHI did not extend to the purchase of the appropriate survey equipment.

---

[105] Dictionary of Fortification, http://civilwarfortifications.com/dictionary/xgb-017.html.

Directly under the topsoil was Locus 1 (Figure 4.29), a deliberate fill layer of relatively hard-packed brown clay with some inclusions of stone and bits of shale.

Figure 4.29. Locus 1. *Left*: view of Locus 1 facing east, excavation in progress. *Right*: view of Locus 1 facing south, nearly 0.90 cm deep, where excavation stopped because Locus 1 supports Locus 3. Photographs by Jill L. Baker.

Locus 1 was situated directly next to and abutting the stone wall (Loci 2/3), which gave it the appearance of having once been an earthen rampart. The composition of this layer was similar to that of the hard-packed clay layer (Loci 3/5) found in the 2011 season. Locus 1 measured 1 meter wide and 1 meter long, north to south on the western side, and almost 2 meters long, north to south on the eastern side. In other words this locus reflects the angle at which it communicates with Loci 2/3 and the north-south orientation of the gridlines. Locus 1 was excavated to a depth of 0.90 cm to almost 1 meter, although it remains open and was not completely excavated because it provides support for the northern wall (Locus 3), most of which remains *in situ*. Discovered in this locus were one rim of a teacup or bowl (Reg. 114), fragments of charcoal (Reg. 115), and some animal bone fragments (Reg. 116).

Part of Locus 2 was located directly under the topsoil and directly south of Locus 1, next to and touching it (Figure 4.30). Locus 2 represents the portion of the wall that has fallen, and while it is unconventional to assign a locus number to wall fall, it was necessary to differentiate between the fallen portion of the wall and that which has remained relatively intact (Locus 3).

This exceptional assignment is important because the intact portion of the wall was largely protected by Locus 1, the earthen rampart, and the fallen portion of wall (Locus 2), which likely extended vertically above Locus 1. This relatively intact section of the wall provided the valuable data sought in this season. Locus 2 stretches in roughly a northwest to southeast direction at approximately a 40-degree angle to the grid line, which is oriented to magnetic north.

Figure 4.30. View of Probe 2, looking north. *Left*: the probe before extending it to 3.50 m. Locus 2, fallen wall, visible in the foreground. *Right*: the probe, extended and partially excavated, revealing the falling action of the wall in the foreground. Photographs by Jill L. Baker.

The stones that comprised Locus 2 included unhewn fieldstones and roughly hewn stones. The stones that appeared to have been worked were generally sandstone and shale, sedimentary rocks that were easily manipulated. The measurement of the stones ranged from fist-sized to ca. 0.50 cm long by ca. 0.40 cm wide by ca. 0.25 cm deep.

Several of these worked stones were found among the wall-fall and were distinguished by what appeared to be fashioned right angles, presumably to accept and direct wooden supports and/or beams. Also found within the wall-fall were several bricks, some of which appeared to have been deliberately modified (Figure 4.31).

Figure 4.31. Possible worked stones that may have been used as architectural elements. *Left*: worked (?) stone in situ. *Right*: collection of some of the stones that may have been fashioned to accept wooden posts or beams. Photographs by Jill L. Baker.

Within the wall-fall was a void that provided access to excavate the sub-flooring. There is little doubt that numerous voids were created when the wall collapsed, but this particular one was much larger and extended with greater regularity than those previously encountered. Presumably whatever once occupied the void has long since disintegrated, but may perhaps represent a wooden beam or architectural element, which would correspond with the presumed worked stones found in the same area.

Locus 3 was located directly under Locus 2 and next to and touching Locus 1 (Figures 4.30, 4.32). This locus represents the relatively undisturbed portion of the northern wall. Locus 3 extends in roughly a northwest to southeast direction at approximately a 40-degree angle to the grid line, which is oriented to magnetic north. Since Locus 3 extends into both the eastern and western balks, its length remains unknown. However, its width is approximately 1.0 to 1.10–1.20 meters wide (ca. 3–3.9 feet wide).

To preserve the wall's integrity, only the uppermost course was exposed, leaving its full extant remains unexcavated, precluding any accurate measurement of its actual height and length for the present. The exposed section of the wall was comprised largely of unhewn fieldstone with some hewn stones ranging in dimension from fist-sized to approximately 0.50 cm by 0.40 cm.

During the construction of the blockhouse, numerous small stones were probably utilized as leveling material, and there was no indication of the use of mortar.

Figure 4.32. Locus 3 *in situ* northern exterior wall. *Left*: view to the south of the wall. *Right*: view to the north of Locus 3, the wall, in the background, and the fallen wall in the foreground. Photographs by Jill L. Baker.

Locus 4 was a soil layer which was located directly south of Locus 3 and under Locus 2 in the probe extension, and consisted of dark black, loose, silty soil, which may represent a natural post-collapse fill layer or more likely a deliberate fill layer that comprised part of the sub-floor fill. There were inclusions of stone and fragments of shale. This layer was approximately 0.3–0.5 cm deep and measured approximately 0.50 cm north to south and approximately 0.80 cm east to west, roughly approximating the size of the void. Glass shards and brick fragments were also found within this layer.

Locus 4 immediately gave way to Locus 5 (Figure 4.33), a thick hard-packed black soil layer with dense inclusions of shale. This hard-packed shale layer has no parallel from the 2011 probe. Perhaps this very solid layer functioned as additional leveling due to the original topography, and/or as a very solid sub-floor fill intended to provide stability to the exterior wall and interior floor joists above which the wooden floor would have spanned, or as an earthen floor to accommodate the housing of animals prior to laying a wooden floor. Given the limited extent of the probe and having left much of the wall-fall *in situ*, for the purpose of preservation, it was not possible to ascertain the way in which this layer (Locus 5) communicates with the wall (Locus 3). Future excavation will have to determine how this layer was associated with the exterior stone foundation/wall.

As Locus 5 was excavated, the abovementioned void or space observed among the wall-fall extended into the southwestern balk. Associated with this layer and extending into the balk was what appeared to be a worked stone (Figure 4.33, Right) and next to it a modified brick. The exposed portion of the stone that protrudes from the balk appears to have been squared off, presumably to cradle a wooden beam. The visible portion of the stone measures approximately 0.42 cm lengthwise (north to south), 0.17 cm wide (east to west) and 0.10–0.15 cm deep. The southwestern corner of the probe extension also yielded architectural fragments. This layer was approximately 1.0–1.20 meters below the surface (measured from the southwestern balk). Found within this layer were ceramic fragments (rims and parts of the bodies) of cups, bowls, and/or saucers (Regs. 117, 118, 119, 120, 121), a stone tool (Reg. 122), the base and part of the body of a glass bottle (Reg. 124), several shards of flat glass (Reg. 125), an animal bone (Reg. 126), a brick, and multiple flat pieces of shale.

Figure 4.33. Probe extension, south of Loci 2 and 3. *Left*: Locus 5, hard packed with inclusions of shale. *Middle*: Locus 6, yellowish-brown hard-packed clay. *Right*: Possible worked stone that functioned as an architectural element, next to a brick. Photographs by Jill L. Baker.

Locus 6 (Figure 4.33) was located directly below Locus 5. This layer consisted of hard-packed yellowish-brown clay with some pebble inclusions. This layer was similar to Loci 3/5 from the 2011 probe. This layer was approximately 3 cm thick. It measured ca. 0.50 cm north to south and ca. 0.80 cm east to west. No artifacts were discovered in this layer.

The final layer, Locus 7, probably represents virgin soil. It was loose to hard-packed with some shale, pebble, and stone inclusions. There were also some sandy and sandstone lenses. This layer was excavated only a few centimeters deep due to lack of time.

It is reasonable to conclude that the section of wall and foundation uncovered during AHI's Third Preliminary Season of the Greensboro

Blockhouse Project appears to be more substantial than those of the non-military structures cited above. By excavating a 1 x 3.5-meter probe across the middle of the northern wall, the defensive nature of the Greensboro structure has been revealed. Although the majority of the foundation and wall remains unexcavated, a conservative measurement of the top course suggests it may be approximately 1 meter (3.28 feet). It appears that the foundation was set solidly into the ground and constructed of fieldstones without the use of mortar. Some stones appear to have been worked, having been flattened or fashioned with right angles. These worked stones may have served as structural elements.

Based on the number of stones found in and around the probe, it is more likely the foundation supported a stone rather than a wooden first-floor wall. Given that blockhouses were usually hastily built structures built of materials that were abundant locally and easily obtained, it may have been more expedient to construct the first-floor walls of stone (or the majority of it), which may have been more readily available than timber, reserving the perhaps more scarce timber for the flooring and the second-story frame, walls, floor, and roof. In this case, the worked stones would likely have been positioned on the upper portions of the wall and fashioned to accept the frame and beams to support the second story. This possibly explains why these elements were discovered in the southwestern portions of the probe (toward the center of the building) rather than adjacent to the intact section of the wall. If this is the case in this instance, it may be assumed that the wall extended vertically, well above ground level.

Outside of the structure, an earthen rampart (Locus 1) appears to have abutted the exterior walls. It remains unclear at this point whether this rampart was structural, intended to strengthen the foundation and wall, or defensive, or both. However, it is clear that the earthen rampart was intentionally constructed given that the composition of the soil, brownish, hard-packed clay, was similar to the hard-packed clay layer (Loci 3/5) discovered in the 2011 season, and assumed then to be a sub-floor leveling layer. Additionally the condition of the soil found in the rampart is vastly different from that found south of the wall or inside the building. Accordingly, the rampart, together with a first floor stone wall, would have ensured a substantial defensive structure.

The northern extent of the 2012 probe begins within the earthen fill/rampart and extends southward across the wall and wall-fall to just inside the building. The southern extent of the probe likely fell to a position just south of (inside of) the northern wall, into a void that may have existed between the wall and a foundation for a floor joist, if it is assumed that there was a wooden floor (cf. Figs. 4.33 and 4.34). However, to date, not one splinter of wood, no negative impressions of wood, nor the stains of rotted wood have been found in either the surface survey or the two probes in the course of three years of excavation. A few charcoal fragments were found, possibly representing burned wood, but this determination awaits further analysis.

Directly below the fallen stones (many of which were left *in situ*) were two layers, possibly sub-floor fill. Locus 6, the lowest layer, was probably a leveling fill. In 2011, a similar layer (Loci 3/5) was identified, suggesting that it extended across the entire site, creating a level surface upon which to build the structure. On top of that, Locus 5 from the 2012 probe was a thicker, more solid layer, which probably provided further support for the stone foundations. This layer may be associated with Loci 2/4 from 2011, but the composition of these two layers differs in that Locus 5 contains much more shale. One can speculate that the original floor was earthen to accommodate animals housed inside the structure, and the timber flooring was a later addition. Found within the deliberate fill layers (Loci 4 and 5), directly south of the wall and wall-fall, were numerous ceramic, glass, and brick fragments, a stone tool and animal bones. It may be assumed these represent occupational debris. In marked contrast to the 2011 probe, in which nails were ubiquitous, not a single nail was discovered in the 2012 probe.

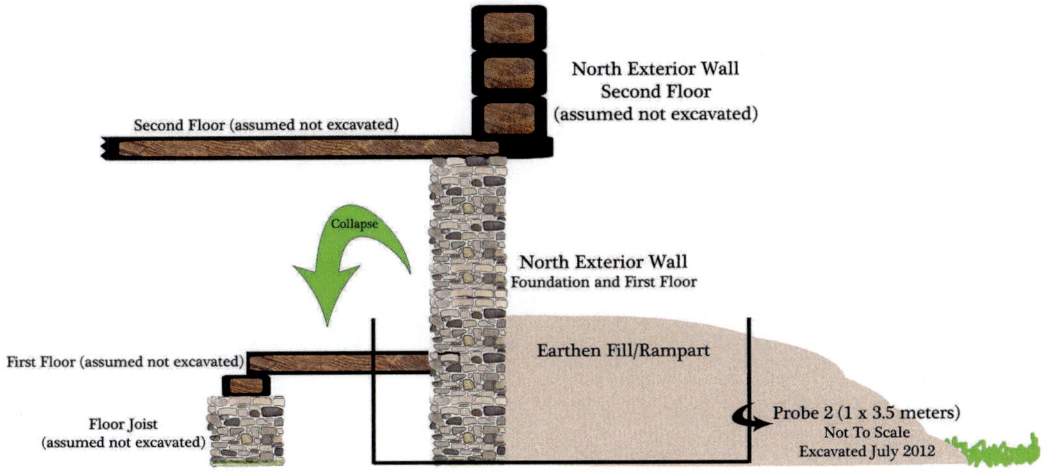

Figure 4.34. Re-creation of the northern exterior wall of the collapsed structure. The weight of the earthen fill/rampart likely caused the stone walls to collapse inward toward the center of the building. Not drawn to scale. Graphic by Jill L. Baker.

Judging from the falling action of where the stones lay, perhaps the weight of the exterior earthwork caused the stone walls to collapse inward, or the walls were deliberately pushed inward rather than allowed to fall randomly (Figure 4.34). Presumably, there was no counter balance inside the structure to offset the weight of the exterior earthen rampart. Thus, the stone walls would have weakened under the weight of the rampart and from the natural movement of the ground due to freezing and thawing, as well as erosion from rain over a period of more than two centuries. When considered together, a distinct correlation may be observed between the stone joist revealed by the 2011 probe and the stone wall and its associated fall from the 2012 probe in that they are parallel to one another. They both extend in a northwest to southeast direction at a roughly 40-degree angle to the gridline (Figure 4.35), and they measure 5 meters (ca. 49.21 feet) apart.

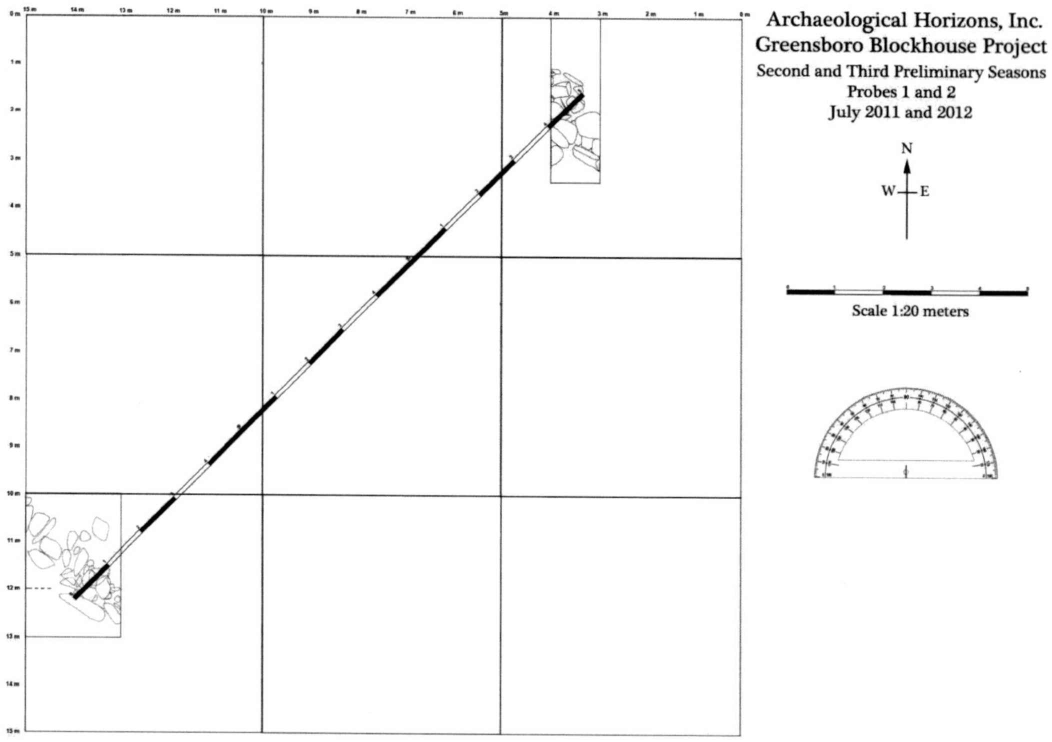

Figure 4.35. Probes 1 and 2 (2011 and 2012 respectively). The two stone features revealed by these probes are parallel and measure some 5 meters (49.21 feet) apart.

It is worth noting the similarities, whether by design or coincidence, that the Greensboro blockhouse architectural remains share with those of the proposed site of the Walden blockhouse (Figure 3.21, Chapter 3 (Site No. VT-CA-24)), the only other extant site of the four blockhouses that were known to be constructed the length of the Bayley-Hazen Military Road.

The northern wall/foundation of the proposed Walden blockhouse measured approximately 74 feet long and was oriented east to west. The eastern wall/foundation measured 38 feet and was oriented north to south (Figure 3.21, Chapter 3. See also Haslam 1997:8). The entire building is oriented northwest (30 degrees) to southeast (ca. 210 degrees). The northern wall appears to have been constructed as a single continuous wall, with the entrance probably located on the building's southwestern elevation. The visible sections of the foundation walls appear to have been dry-laid fieldstone. A well was located within the confines of the structure in the partially enclosed area in the southwestern corner of the proposed site. Based on our observation and archaeological investigation, it was constructed with dry-laid slate capped with two slate covering stones, and although it is now filled with rocks, it is judged to have been approximately 20 feet deep. It should also be noted that the site of the Walden blockhouse is considerably less well preserved than that of the Greensboro blockhouse.

The Fourth Preliminary Excavation Season, 2015

Operating under the overall project goals, the focus for the 2015 preliminary excavation season was to investigate an area presumed to be that of the fireplace/chimney and the western wall that extends in a north to south direction. In general, the fireplace/chimney of a stand-alone blockhouse would have been located in the center of the building or to one side. From Jonathan Elkins' description of the Peacham, Walden, and Greensboro blockhouses, the fireplace/chimney was likely located centrally. From the data collected in the 2010 survey, a large worked stone lying toward the top of the stone rubble may be a mantel or a hearthstone, and partial bricks may represent the remnant of a chimney or fireplace. It is assumed this large stone marks the fireplace platform and chimney. So as not to disturb what is likely a very significant element of the blockhouse, we chose to excavate Square NE4, which theoretically would have allowed us to peer between the fireplace platform and the western exterior north-south wall. However, as shall be seen, the site had different intentions.

Using the grid that was established in 2011, Square NE4 is located between the 5 and 10-meter marks along the north-south center grid line and the 10 and 15-meter marks on the northern east-west grid line (Appendix 2). The central north-south grid line also roughly corresponds with the western wall that extends from north to south. Once the area had been prepared and the parameters of the square established, excavation of Square NE4 commenced. As the excavation proceeded, it became clear that part of the square should remain unexcavated until a time when proper restoration of the fireplace/chimneystack could be undertaken and when proper survey and recording equipment could document it. Therefore, the northeastern portion of the square remains unexcavated.

Once the vegetation was cleared and the topsoil removed, we investigated the layers and features below. The topsoil varied in depth from 1 cm around and among the rocks to 10 cm where there were no rocks. The elevation of the topsoil was 465 meters above sea level according to an application, GPS Status, on a Samsung Galaxy S4.

Directly under the topsoil was Locus 1 (Figures 4.36, 4.37), a fill layer in the northwest corner of Square NE4 directly next to and contiguous with Locus 2. The composition of this layer consisted mainly of loose brown soil with several floating rocks (not associated with another feature). Also among this layer were some brick fragments. This layer was excavated to a depth of ca. 10 cm, and measured ca. 0.30 to ca. 0.50 m north south and ca. 0.75 m east west. This locus was wedged in the corner between the north and west balks and Locus 2. Locus 1 was located directly above Locus 11.

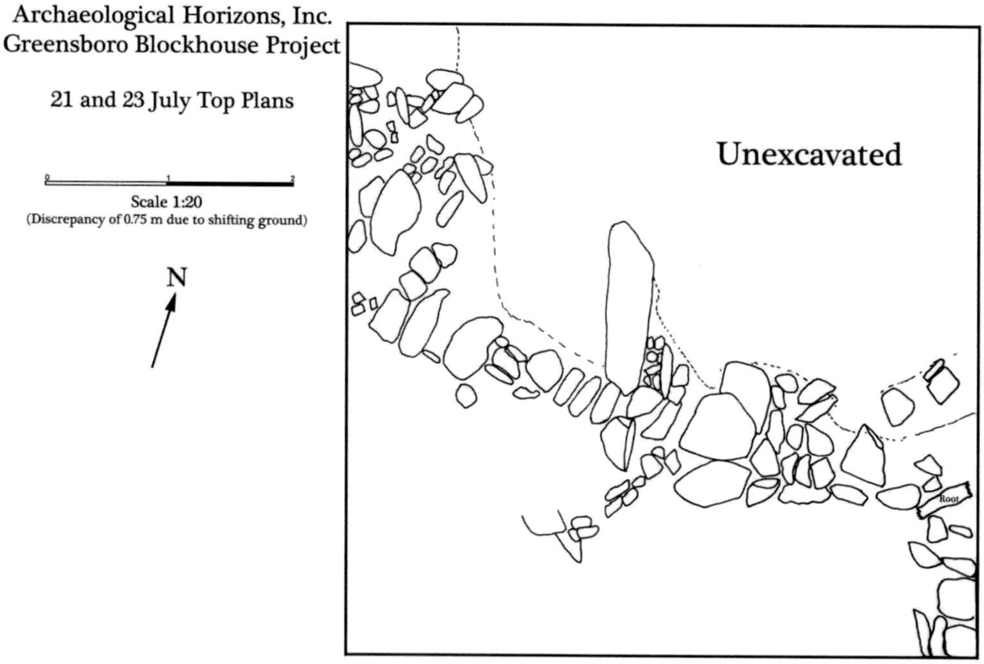

Figure 4.36. Top Plan #1, July 21, 2015. Top plan by Jill L. Baker.

Figure 4.37. Loci 1 and 11. *Left*: Locus 1 under excavation. Middle: Locus 1 removed with Locus 11 beginning. *Right*: Locus 1 in section. Photographs by Jill L. Baker.

Locus 2 was a rock feature directly under topsoil next to and contiguous with Loci 1 and 3 (Figures 4.37, 4.38). Locus 2 was located at the northern part of the west balk between approximately the 5.25–5.30 meter and 7.25-meter marks, extending diagonally across the square in a southeasterly direction, meeting and merging with the unexcavated area. Locus 2 formed a relatively straight line running in a southeast-northwest direction at roughly a 70-degree angle with the north-south grid line. Within this feature were several nails (Regs. 130, 141, 188), a ceramic rim fragment (Reg. 142), and a ceramic body fragment (Reg. 143). Some bits of charcoal and partial bricks (Regs. 197–202) were also discovered among these rocks. This rock feature was only two courses, three at most, and measured approximately 0.25–0.50 m deep, 2.0–2.50 m east to west, and 1.5–2.0 meters north to south. The rocks that comprised this feature were mostly unworked fieldstones set in place using a dry masonry technique. The rocks used in this feature ranged from fist-sized to larger, relatively flat stones. One measured 64 x 28 cm and contained marks suggesting its use as a structural element (this stone, Reg. 203, discussed below). Locus 2 was situated directly on top of Locus 9. Although most were unworked fieldstones, some rocks appear to have been worked. Locus 2 appears to have formed a relatively straight and level line.

Figure 4.38. *Left*: Loci 2, 3, 4. *Right*: Loci 2 and 9. Photographs by Jill L. Baker.

Locus 3, a rock feature, was located next to Locus 2 and above Locus 9 (Figure 4.38). Locus 3 appears to have been at a slightly lower elevation than Locus 2, approximately 0.20–0.25 m lower. Locus 3 extended from the western balk in a southeast northwest direction diagonally across the square. It too was at a 70-degree angle to the gridline. This rock feature appears to have been only one course deep; the width of the large rocks that comprised it measured approximately 0.20–0.25 m deep by 0.50–0.75 m wide. The rocks appear to have been unworked fieldstone, mostly of an elongated, somewhat flat shape ca. 0.25–0.50 m long. This rock feature, likely a sill, formed a relatively level surface. Discovered among the rocks were several artifacts: three nails (Regs. 129, 131, 133), a ceramic fragment (Reg. 132), and some brick fragments.

Locus 4 (Figure 4.38) was located in the southwestern portion of the square, just below topsoil, next to Locus 3 and contiguous with Locus 6, which extends eastward from the 7.50-meter mark. This was a loose-packed black soil layer that included wall fall as well as nails (Regs. 134, 135, 136, 137) and glass (Reg. 138). Locus 4 measured ca. 0.20–0.25 m deep and was situated over Locus 9.

Locus 5 is a rock feature just below topsoil (Figure 4.39). It extends from the 2.50-meter mark at the center of the square eastward toward the eastern balk of Square NE4. Locus 5 represents the continuation of the rock feature Locus 2 and possibly Locus 3. Some wall fall was associated with Locus 5 as well as Locus 3. No artifacts were discovered among this rock feature, as it remains unexcavated, but there were some brick fragments. The stones were mostly unworked fieldstone; however, one appeared to have been worked, possibly functioning as an architectural element (Figure 4.39, Right).

Figure 4.39. *Left*: Loci 5, 6, 7/8, and 10. *Right*: possible architectural stone. Photographs by Jill L. Baker.

Locus 6 (Figure 4.39) was a soil layer that extended under Locus 5 to the southeastern portion of the square. It began in the middle of Square NE4 at the 2.50-meter mark and was contiguous with Locus 4 as well as the wall-fall that was possibly associated with the Loci 5 and 3 extensions. Locus 6 was a black, loose-packed soil layer that measured approximately 1.0 m wide and 0.20–0.25 cm deep. Numerous artifacts were associated with this locus, found mostly under/among the wall-fall associated with Locus 5 and extending toward the eastern balk. These artifacts include ceramic fragments (Regs. 139, 175, 176), glass fragments (Regs. 177, 187), some bits of charcoal (Reg. 145) and nails (Regs. 140, 178–183, 185–186).

Loci 7 and 8 are essentially the same fill layer (Figure 4.39). An arbitrary division was made in the field for logistical purposes; here they will be treated as the same, Locus 7/8, which extends from just beyond the 2.5-meter mark on the east-west square line running along the southern balk and into the eastern balk. Locus 7/8 was a loose-packed dark soil fill layer that extended under Loci 5 and 10. Locus 7/8 was approximately 1.0 m wide and 0.20–0.25 cm deep.

Numerous artifacts were discovered in this locus, mostly in the southeastern corner near and under Loci 10 and 5. These artifacts include a hoe head (Reg. 144), nails (Regs. 160–166, 195), ceramic fragments (Regs. 167–172), glass (Reg. 174) and a small round stone (Reg. 173).

Locus 9 was a brown crumbly clay-like fill layer that extended under Loci 2, 4, 6, and 1 (Figure 4.38). This fill layer was also encountered in the 2011 season. As in the 2011 season, Locus 9 is considered a deliberate fill layer to make the building site level. No artifacts were discovered in this layer.

Locus 10A was a gray hard-packed layer with significant shale inclusions. No artifacts were discovered in this layer. A similar layer was encountered in the 2012 season; considered a sterile layer, it may have been used as a stabilizing/leveling layer or as the initial ground floor, or as a leveling layer meant to provide stability and drainage to the lowest foundations of the blockhouse.

Locus 10 was a stone feature located in the southeastern corner of the square (Figure 4.39). This feature extended into the balk and remains largely unexcavated. However, it was also associated with Loci 5 and 7/8 and the numerous artifacts mentioned above, as well as a nail (Reg. 196).

Locus 11 was located in the northwest corner directly under Locus 1 and the unexcavated portion of the square (Figure 4.40). Locus 11 was a pit that contained ash and charcoal as well as numerous artifacts and bricks, whole and partial. Unfortunately, this locus extended into the northern balk and remains largely unexcavated. Locus 11 measured approximately 10–13 cm deep and 20–25 cm wide.

Numerous artifacts were discovered in this pit. These include ceramic fragments representing at least 10 vessels (Regs. 146–154), a metal fragment (Reg. 155), glass fragments (Reg. 156), 3 nails (Regs. 157–159) and six animal bones (Regs. 189–194).

Figure 4.40. Locus 11 in section. Photography by Jill L. Baker.

By way of review, the goal of the Fourth Preliminary Excavation Season for the Greensboro Blockhouse Project was to investigate the area between the western wall that extends in a north-south direction and the presumed fireplace platform and chimney. To accomplish this task the dig team excavated a significant portion of Square NE4; however, the portion of Sq. NE4 that contains the majority of the presumed fireplace/chimney remains unexcavated until such time as proper post-excavation preservation of significant features can be undertaken. Identification of the fireplace is important for determining the function of this structure. Blockhouse architecture frequently located the fireplace in the center of the building. Several contemporary blockhouses that utilized a central fireplace/chimney include the Lacolle Mills Blockhouse in Ontario, Canada, the Fort Halifax Blockhouse in Winslow, Maine, and the Fort Kent Blockhouse in Fort Kent, Maine. This season's goal was to excavate an area that would allow us to see both the outer wall and the fireplace in section; however, there was an unexpected surprise.

Based on the 2010 surface survey, a large worked stone, presumably a hearthstone or a mantelpiece, was located in what appears to be the center of the site. This stone measures approximately 1.39 m long x 0.30–0.40 cm wide on two sides. Much of the stone remains unexcavated and partially buried. Nevertheless, the size and shape of this stone was significant. Therefore this season we chose to excavate part of Square NE4 hoping to gain a better understanding of this area.

Once the square lines were established and digging commenced, it became clear that we were inside the structure but not in an area that would expose the western wall. What was encountered was a rock feature, Locus 2, which appeared to emerge from the western balk extending to the presumed fireplace platform. Next to and south of Locus 2 was Locus 3, which may have been either wall fall or a stone sill to carry wooden beams, which is more likely (Figure 4.41). A clear line could be detected between the two loci, which extended across the square, forming a sort of ledge. These two stone features are at the same 70-degree angle to the balk line as the rock feature, the floor joist/wall from 2011, and the northern wall excavated in 2012. Therefore, it is likely that Locus 3 functioned as a sill on which timber framing for flooring could have been set.

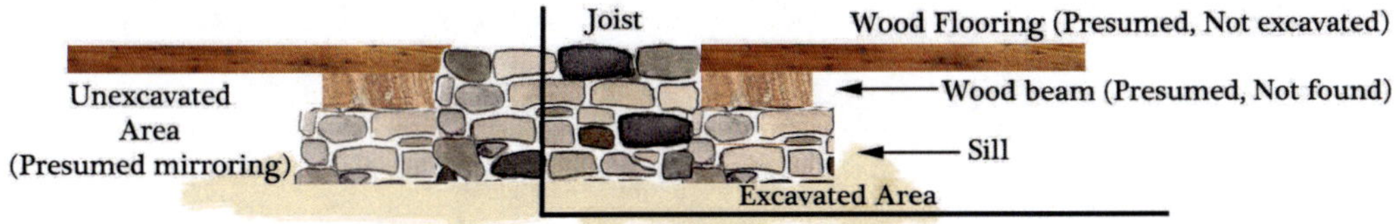

Figure 4.41. Possible rendering of the excavated area with the joist (Locus 2), sill (Locus 3), and presumed wooden beam and flooring. No wood was discovered during the 2015 excavation season. Graphic by Jill L. Baker.

Beyond Loci 2 and 3, as the floating rocks, presumed wall fall, were removed, among and below them were several somewhat concentrated deposits of nails as well as ceramic and glass fragments. These deposits were located south of the 'sill' in the west, center, and east portions of the square (Figure 4.41).

The nails were in repetitive clusters in combinations of large, medium, and small sizes (Figure 4.42) and ranged in length from ca. 9.5 cm to 3.5 cm. These clusters included as many as 8 nails. In all, 30 nails were recovered. Of these, 20 (66.6%) were bent or broken, and only 10 remained straight. All but 2 of the nails had square shafts with flat heads of rectangular shape measuring approximately 1.0 cm x 0.5 to 0.8 cm.

Figure 4.42. Approximate general locations where the largest concentrations of nails, ceramic, and glass were discovered. Photography by Jill L. Baker.

Two nails had round heads with slightly square shafts just below the head tapering quickly to a round shaft (Regs. 165, 180). Although numerous nails were discovered, not one splinter of wood was recovered, which has been consistent for all dig seasons. The high percentage of bent nails suggests that they were removed from the wood they once bound; presumably the wood was salvaged and removed while the nails were left behind among the rubble. Bent nails may also represent clenching.[106]

[106] The tip of the nail is hammered flush with the wood to secure, or clench, the wood pieces together.

Once the sill was removed, a clear line of stones marked the joist/fireplace platform as it stretched across the square (Figure 4.42). The position of the nail clusters just next to (south of) the sill suggests that Locus 3 once supported a wooden beam that extended diagonally across the square to support timber beams for flooring.

Among the rocks that comprised Loci 2 and 3 and among the wall fall south of them, two rocks bear mentioning (Figure 4.43).

Figure 4.43. Square NE4, looking north. *Left*: two *in situ* rocks highlighted in hot pink. *Top Right*: stone with markings; *Bottom Right*: possible socket. Photographs by Jill L. Baker.

Referred to in the field as the "petroglyph stone," the upper rock appears *in situ* as part of the joist that extends from the western balk to the 'fireplace platform' (Figure 4.43, Reg. 203). This rock contained a number of impressions that suggest it was used as a support for posts and/or joists, and possibly it contains intentional incised markings. Some of these marks appear as a recurring chevron-like pattern in several places on the stone. On the front side there appears an incised, elongated, horizontal Y-shaped character, possibly a mason's mark. Two incised oval or round shapes on top of the rock may mark the placement of posts. The second lower *in situ* rock highlighted in Figure 4.43 appears to be rounded with one flat side and a sort of concave dimple impression on the other side. This rock may have functioned as a socket (for a door) or as the base of a post or a hammer or grinding stone.

The ceramic fragments included rims, bases, and body shards. The majority of the wares were white-glazed, some with decoration such as that depicted on a blue scalloped rim fragment (Reg. 139), a body fragment with a floral decoration (Reg. 132), a body fragment decorated with a pineapple (Reg. 168), two joining rim fragments with a blue geometric pattern (Reg. 169), a black glazed body fragment (Reg. 170), multiple non-joining fragments of a red glazed vessel (Reg. 154), and two rim fragments of a cup (?) with an interior horizontal red band just below the rim, one with an external floral decoration (Regs. 149, 176). The glass fragments were usually small, and all were flat with a greenish tint. The ceramics and glass were generally discovered among the nail clusters.

Once the western portion of the floor joist (Locus 2) was removed, it was clear that it abuts what is presumed to be the fireplace stack, which remains unexcavated. Below Locus 2 was a crumbly brownish clay-like layer, probably a deliberate fill layer used to make the area level. This same layer was encountered in the 2011 season. The northern part of Locus 2 extended into the western balk leaving a small portion of the northwestern corner of the square free for excavation. This was Locus 1, a fill layer of dark soil. Once Locus 1 had been removed, a pit comprised of brownish-gray material that included charcoal and ash was visible (Locus 11). Unfortunately, this pit extends into the northern balk and remains only partially excavated, its full extent and contents unknown. Nevertheless, a surprising number of artifacts were recovered from this small pit: fragments representing as many as 11 ceramic vessels (Regs. 146–154), three nails (Regs. 157–159), six animal bones (Regs. 189–194), one glass fragment (Reg. 156), and one metal fragment (Reg. 155). Among the ceramic fragments were numerous joining rim fragments of a flare-rimmed plate or shallow bowl (Figure 5.46 left, Chapter 5). Presumably the rest of this vessel remains buried. Several bricks also remain *in situ* in the section of the north balk.

Extending into the southeastern balk were more stones, Locus 10 (Figure 4.39). It is likely that this feature represents the eastward extension of a joist that joined the fireplace platform to the eastern wall of the structure. This is speculative, as this area remains unexcavated. However, this theory can be supported by the cluster of nails and ceramic fragments discovered in the area immediately next to Locus 10 and the 'sill' (Locus 3). Among the rock features some bricks were discovered (Figure 4.44).

Figure 4.44. Bricks found during excavation, partial bricks that appear to have been modified in antiquity. Photographs by Jill L. Baker.

Most of the bricks were partial and appear to have been deliberately cut, not necessarily broken due to collapse. These were discovered in random order, and their position did not form any noticeable patterns except for those remaining in section in the northwestern balk. These bricks appeared to have been mold-made, string/wire cut, and fired. Additionally, none of these bricks contained any traces of cement.

In summary, the 2015 excavation season exposed what appears to be a floor joist that extends from the western wall to the fireplace platform at the center of the structure, and the beginning of another joist that extends toward the fireplace platform from what is presumed to be the eastern wall. Our initial expectation was that the sub-floor foundation and fireplace platform would look much like those of the Lacolle Mills Blockhouse (1781–1838), which appears to have had a central sub-floor fireplace platform and lip to receive wooden beams and flooring, but without the aid of intervening joists (Figure 2.7, Chapter 2).

The Greensboro blockhouse, however, appears to have had central joists to support the beams and flooring. This type of sub-floor support system ensures a very stable structure. Based on the findings from the 2011, 2012, and 2015 preliminary seasons, the stone features from Probes 1 and 2 may represent outer walls. The wall found in Probe 2 certainly represents the outer northern wall, and its height and width are consistent with a fortification structure. The wall found in Probe 1 parallels that of Probe 2 and may represent the southern outer wall (Figure 4.45).

Between the two is the structural feature excavated in the current 2015 season. This feature likely represents the southern portion of the fireplace platform and floor joists extending from it to the east and west walls, both of which remain unexcavated. It would be reasonable to expect the fireplace platform to extend northward into Square NE5 and for it to have an adjoining sill and joists to receive a timber beam to support timber flooring. This would then leave approximately 4–4.5 meters for timber flooring to span between the fireplace platform/sill/joists and the outer walls on each side. The overall measurement appears to be ca. 15 meters or approximately 49.212 feet across. If the blockhouse were square, this would give an estimated total measurement of almost 50 x 50 feet. It may also be that the original flooring consisted of the hard-packed shale layer found in the 2012 season in Probe 2, and the stone joist/sill found in the 2015 season was added later to accommodate wood flooring.

The location of the fireplace and joist(s) excavated in the 2015 season appear to be roughly in the middle of the structure when measured between the joist from Probe 1 (2011) and the outer wall found in Probe 2 (2012) (Figure 4.45).

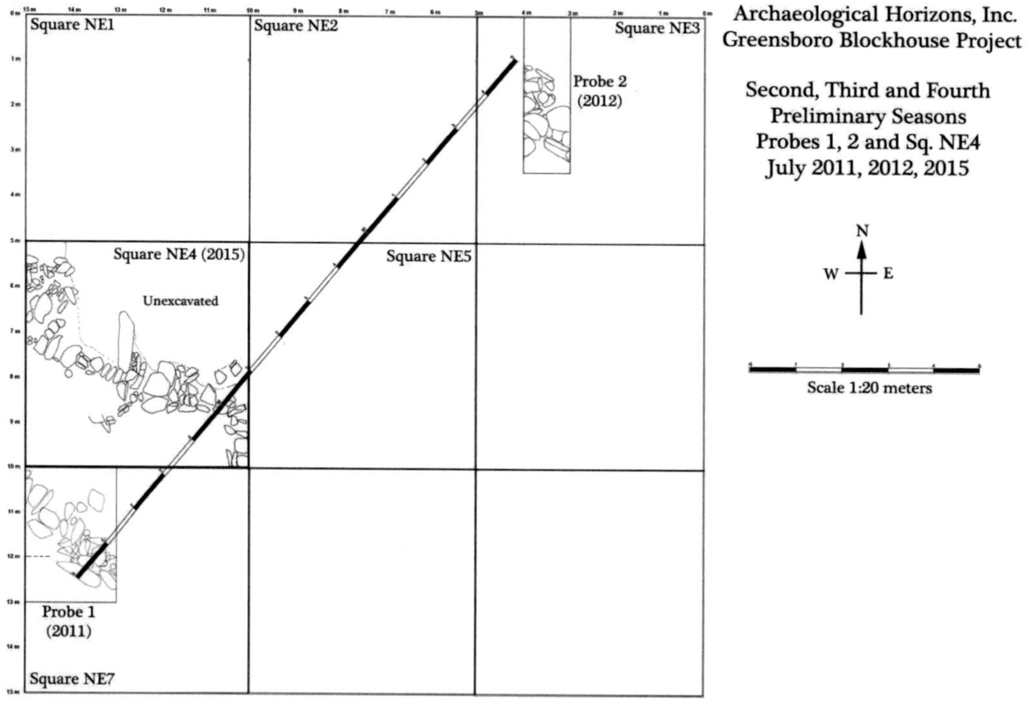

Figure 4.45. The architectural features from the 2011, 2012, and 2015 preliminary seasons placed onto a master top plan within the grid. Top plan by Jill L. Baker.

Although the width of the fireplace/chimney is not known at this time, its width could be assumed to be at least approximately three meters, which would place the fireplace/chimney in the middle of the building. This layout matches Elkins' description of the construction of the Walden blockhouse; the size, construction technique, and layout would no doubt have been similar at the Greensboro blockhouse. A large stone in the middle of the Walden blockhouse ruins may also be a hearthstone, which would correspond with the large stone found in 2010 and 2015. According to Elkins' description, the chimney would have been made of fieldstone with beams running "...over all around the house..." (Elkins' Reminiscences). The presence of brick suggests that the fireplace/chimney was originally constructed with some brick, perhaps lining the firebox, or later modifications were made by those who inhabited the structure during the residential phase. Whatever their function, the presence of brick suggests that some were incorporated into the construction in some form, but the majority were perhaps salvaged, just like the timber, for use elsewhere leaving behind only these few partial and whole (2010, 2011) bricks.

Preliminary Interpretation

At present, a very small percentage of the site, 14.44%, has been investigated.[107] To offer firm conclusions about this site based on the data collected from excavation conducted thus far would be premature and inappropriate. However, we may offer preliminary observations and suggestions with the understanding that data produced from subsequent excavation may challenge and change that which is offered here. Our goals have been to identify the true nature of the collapsed structure and to identify the military and residential phases of the structure, assuming it was the Greensboro blockhouse. The data recovered thus far appear to confirm that this was a substantial structure built for purposes other than a residential dwelling, but that also served as such in the ensuing years.

The form and function of architecture work — the typology of a structure — often reveals its function. For example, the architectural designs (forms) of railway stations, churches, and castles generally foretell their purpose (function). Similarly, the architectural design of a blockhouse is unique in that it possesses certain elements and characteristics that portend its function as a military structure. Specifically, there are two aspects of the structure that suggest its purpose as a military blockhouse: the substantial foundation and the location of the fireplace/chimney.

---

[107] This calculation is based on the grid established in 2010. The grid measured 15 x 15 meters. Of that, 32.5 square meters, or 14.44% of the grid, has been excavated thus far.

In the 2012 season we explored a small portion of the northern foundation and wall, Probe 2. The purpose was to assess its size and composition to determine whether the structure was a blockhouse, a farmhouse or some other non-military structure. Based on what we found, it is very likely that the structure was military rather than civilian.

Probe 2 revealed a substantial stone wall strengthened by an outer earthen berm or small rampart that extended approximately 3 meters to the north in a sloping manner. As mentioned above, earthen ramparts have been in use since the Bronze Age in Canaan (ca. 2000 BCE) and remained in use throughout history. Earthen ramparts were utilized to increase the width and height of city walls. This made it difficult for sappers[108] to tunnel under the city wall to get to the city. In this case, the earthen rampart is much smaller, but it provided additional support and strength to the foundation. This would deter sappers from tunneling under the blockhouse and emerging inside it, and from weakening the foundation of the blockhouse to cause it to collapse. The rampart appears to have been made from a yellowish-brown hard-packed clay, which was much different from the surrounding soil. This fill layer abutted the foundation wall, which suggests it was placed there deliberately and purposefully. The fieldstone foundation wall was approximately 1.0 m (ca. 3.28 ft.), a very solid foundation. Many of the fieldstones appear to have been worked, suggesting careful fitting of them into the foundation so that they could receive wooden beams to support the rest of the structure. From inside the structure, the foundation trench was quite deep, approximately 1.0 to 1.20 meters. Such a deep foundation would provide great stability to the base of the structure. Below the foundation wall was a thick substantial shale layer (2012, Locus 5) and a yellowish-brown clay-like layer below that (2012, Locus 6). Locus 6 may have been a leveling layer that likely provided stability for the structure, and Locus 5 may have served as the structure's original surface.

Based on this data the following scenario may be possible. The foundations were set deep into the ground to provide strength, stability, and defense against intruders and artillery. Inside the structure, the ground floor may have been composed of crushed shale and soil, providing a solid floor suitable for housing equipment and animals. At some point, whether at the time of construction or later, some fieldstones were modified to receive timber beams that would support a timber floor and superstructure. The fieldstone walls/floor joists discovered in the 2011 and 2015 seasons and associated nails suggest that timber beams and flooring spanned the structure and were supported by these solid joists. At what point the timber was added, whether at the time of construction

---

[108] Sapping was an ancient technique whereby invading armies would dig a tunnel below city fortification walls with the intention of weakening the wall causing collapse or to gain entrance. See Bentley 1999:17,51-57,104, 114; Ancient Discoveries 2008, and Oleson et al. 2008:685.

or later, cannot be determined at present. What can be said with confidence is that stone and timber worked in concert to provide flooring and presumably a timber superstructure. It is also difficult to determine the height of the walls above the foundation and whether they were made entirely of fieldstone or a combination of fieldstone and timber.

With the excavation of Square NE4 in 2015, we investigated an area assumed to be that of the fireplace/chimney. We had hoped to find a sub-floor platform in the center of the structure to support the fireplace/chimney and a void between that platform and the outer wall. However, we did not find such a neat configuration. Instead what we discovered was likely another fieldstone floor joist/sill (Loci 2/3) extending from the outer western wall to the fireplace/chimney platform and another joist likely extending from the outer eastern wall inward to the platform, though the eastern extent remains largely unexcavated. The large quantity of nails and the variety of their size and shapes suggests the presence of timber beams and flooring that were once supported by these stone joists. Nails were found in clusters and in predictable quantity and size in fill south of the stone joist/sill. The edge of this joist (Loci 2/3) was at a 70-degree angle to the grid line, the same angle of the foundation wall in Probe 2 (Locus 3) and of Locus 1 in Probe 1 (excavated in 2011). Furthermore, the nails discovered in Probe 1 in 2011 were north of Locus 1, presumed to be a floor joist. These nails were also found in clusters and in predictable patterns of number and size. The area that was north of Locus 1 in Probe 1 and south of Loci 2/3 in Square NE4 would have been the sub-floor surface and fill below the timber flooring of the building. In other words, timber beams and flooring would have spanned between Loci 2/3 in Square NE4 and Locus 1 in Probe 1. The distance between the edges of these fieldstone joists would have been approximately 3.0 to 3.5 meters.

In the area between the two joists, 53 nails were discovered in association with 2011-Locus 1 and 30 in association with 2015-Loci 2/3. Thus, 83 nails were discovered between the joists. Of that number, 54 (65%) were bent or broken. Additionally, not one splinter of wood has been discovered during any of the exploratory seasons. This, coupled with the bent nails, suggests that the timber from the flooring and the presumed superstructure was removed, leaving behind the nails. It could be that those inhabiting the blockhouse as a residence were aware of its imminent collapse. Perhaps the salvageable materials, such as timber, were removed prior to its demise and reused elsewhere. Then, once the usable materials were removed, the remaining portion of the building was demolished in a controlled manner. Evidence for the structure's controlled collapse is demonstrated by the fact that the stone walls appear to have all collapsed inward toward the center of the building rather than in a chaotic manner, as is the case with structures that collapse on their own. In 18th century America, iron nails were in short supply and high demand. One way to obtain nails was to burn buildings to the ground and retrieve the nails from the ashes.[109] In this case, however, it appears that the timber was more valuable than the nails.

In conclusion, the four preliminary excavation seasons conducted by Archaeological Horizons, Inc. at the presumed site of the Greensboro blockhouse have produced tantalizing clues as to the nature of the collapsed structure, and have offered fresh and exciting perspectives on the interpretation of this site. The ceramics and nails recovered in the course of excavation suggest the structure was built and inhabited in the late 18th and early 19th centuries. The ceramics found within the structure are similar to those commonly found at contemporaneous colonial period sites, most of which were likely imported from England. The majority of the artifacts appear to represent the residential rather than the military phase of occupation. That few, if any, military artifacts have been discovered in the course of four seasons of excavation should not be surprising given that excavation has focused primarily on the interior of the structure, and the residential phase lasted longer and also represents the final phase of its use. A similar phasing sequence of the occupation and use of the Ft. Pitt Blockhouse in Pittsburgh, Pennsylvania, produced similar results. The substantial stone foundation, earthen rampart, and first floor stone wall are evidence of the structure's original military purpose and use. An unidentifiable metal object was discovered during the 2015 season, together with numerous quartz flakes, which could be the debitage that results when fashioning stone tools from a core stone. These and other stone flakes/tools found in previous seasons may belong to the earliest phases of occupation, and may have been used by native peoples or soldiers after the war ended.

Considerable and valuable data have been produced and new knowledge gained during the four seasons of excavation. Simultaneously, as is the nature of archaeological investigation, numerous questions have arisen. Why has no trace of timber been discovered? Were the second story and roofing secured by wooden pegs? Was the original flooring made from a hard-packed clay and shale mix, and a wooden first floor added later? Was there a stockade fence? If so, where was it? Did additional buildings house animals and ammunition? Was there a well, and if so, where was it located relative to the structure? These and other questions remain to be addressed and resolved until further excavation of the site can be undertaken. Nevertheless, for the present, unless substantial evidence to the contrary is produced, the knowledge gained based on the work accomplished to date is more than adequate to presume that this site is that of the Greensboro blockhouse constructed during the Revolutionary War.

---

109 C. E. Peterson. 1950 Burning Buildings for Nails. American Notes. *Journal of the Society of Architectural Historians.* 9, No. 3. P. 23. G. LeFever, 2008 Forged and Cut Iron Nails. Early American Life.

Although the Bayley-Hazen Military Road and its four blockhouses are considered by some to have been a military ruse or even part of a failed campaign into southern Canada, they comprise a little known chapter in the history of the emerging United States of America. Considerable energy and resources were devoted to this project by a significant group of people, including General Washington, the Second Continental Congress, General Bayley, Colonel Hazen, the laborers who constructed the road and blockhouses, and the soldiers assigned to patrol the area. This project, historical and modern, highlights aspects of Washington's strategic planning and the ambitions (some self-serving) of those who promoted the road's construction. Irrespective of whether it was a ruse, a failed campaign, or a manifestation of selfish ambition, the military road and its blockhouses constitute a significant moment in the early history and formation of the nation, and, more importantly, to the early development of the communities of Greensboro, Cabot, Walden, and Peacham in northern Vermont.

Chapter 5

Ceramics, Nails, and Miscellaneous Artifacts

In that architectural features, such as foundations and walls, define a structure's intended purpose, the artifacts found within reflect those who lived within its walls. Artifacts, or the 'stuff' people leave behind, reveal much about socio-economic status, occupation, and life style. Artifacts also provide information regarding domestic and international trade, manufacturing methods, and availability of resources. The artifacts found within the Greensboro structure will provide information about its inhabitants, their economic status, and personal tastes, and should reveal both the military and civilian occupation phases. This chapter will discuss the ceramics, nails, and other material culture found during the course of excavation. As discussed above in Chapter 4, note 2, our original intention was to include a Catalog of Artifacts that would share knowledge of all the artifacts; however, it was cost prohibitive. Instead, this chapter will include photographs and descriptions of the best examples from each category and all artifacts will be mentioned by registration number.[110]

Ceramics

One of the most important artifacts recovered from any archaeological excavation is pottery, also called ceramic. Ceramic was originally a Greek word, κεραμικός (*keramikos*) meaning *of or for pottery.* The earliest reference to the root, *keram*, comes from a Linear B Mycenaean Greece text using the word *keramewe*, or *workers of ceramics.* Ceramics

[110] For those who wish to inquire about the artifacts further, please contact AHI at archaeologicalhorizons@gmail.com.

comprise not just vessels, such as jugs, jars, plates, and bowls, but also objects such as figurines. One of the earliest ceramic objects recovered from antiquity thus far was a figurine of the female fertility deity often referred to as Venus, which dates to ca. 29,000–25,000 BCE and was discovered in 1925 in Moravia. Ceramic vessels and figurines have been extremely helpful in identifying cultures, technology, chronology, and the movement of peoples in ancient and modern times.

When pottery is found in secure archaeological context, vessels are categorized according to type, which includes characteristics such as size, shape, decoration, clay composition, and cultural affinity. Diagnostics include rims, bases, handles, shoulders, and body shape. These typologies and their characteristics can be further categorized according to chronological period. Thus, a ceramic vessel can reveal origin of manufacture, chronological period of use, and the people-group who made it. The origin of manufacture is important because it suggests trade and/or the movement of peoples. The chronological period is important because it can help to date associated architecture and artifacts. Petrographic analysis, the examination of inclusions of a thin section of a ceramic vessel, can accurately determine the original clay source, which can suggest where the vessel was made or to whom the clay was traded and transported.

Since antiquity, after clay is extracted from the ground, it has to be levigated, or washed. Levigation, practiced by both ancient and modern potters, "is the process by which the coarse fraction of a clay, and any large aplastic inclusions it may contain, is separated out from the fine fraction."[111] The clay is washed in a series of pools, tanks, or pits filled with water. The clay is kneaded by hand or foot, which helps to separate the inclusions from the clay. Inclusions can be twigs, stones, or anything other than the clay itself. The clay is then left to settle, another way of removing impurities from the clay. Finer clay is levigated thoroughly, coarser clay less so. Coarser clay was typically used for cooking pots, water jugs, or storage jars. Well-levigated clay was primarily used for tableware such as plates and bowls. The ceramic fragments found during the preliminary seasons of excavation at the proposed Greensboro blockhouse site help to provide a chronological and cultural context for the structure.

The pottery of New England can be divided into four basic categories: red earthenware, stoneware, buff or white body cast, and decorative pottery.[112] Although the potters of New England brought with them the ceramic technique, technology, and decorative traditions of England and Europe, local resources made it very difficult for early potters to reproduce the quality and type of wares they had produced at home overseas. The local

---

[111] Bourriau, Nicholson, Rose 2006:122.
[112] Watkins 1950:1.

clay was coarse and difficult to process, and there was a lack of tin and lead needed to produce finer wares such as porcelain. Nevertheless, potters used the coarse clay to produce and distribute earthenware such as red ware and stoneware, the most commonly used of the locally made ceramics. Finer ceramic wares were imported from England and Europe and included English tin/lead-glazed earthenware, stoneware, Delftware, Staffordshire slipware, Fullham stoneware, salt-glazed stoneware, creamware, sgraffito slipware, Wrotham slipware, English porcelain, shell-edged pearlware, and blue and white Chinese porcelain, to name a few.[113]

Some of these coarse and fine wares were discovered at the Greensboro blockhouse site. The ceramic artifacts recovered from excavation were all fragmentary. In some instances several sherds joined to make a larger fragment; however, no whole ceramic vessels have been discovered thus far. Nevertheless, the fragments that have been recovered offer clues as to the character of those inhabiting the structure and the period of their habitation. These fragments also attest to the residential phase of the structure, as it is not likely that army personnel would have transported or used such refined tableware, though some of the earthenware may have been from the military phase of occupation.

In total, 77 ceramic fragments were discovered representing 45 individual vessels. Whenever individual sherds were joined or clearly of the same vessel, they were all considered to comprise one large fragment. Sherds that did not join or resemble other fragments were considered to represent separate vessels. Further analysis may prove that some non-joining fragments were in fact part of the same vessel. However, at present, of the 77 total fragments, the 45 joined and non-joined fragments are considered to represent independent vessels and will be referred to as '45 fragments'. Ceramic fragments were discovered in all three of the excavation seasons; however, the majority was discovered in Probe 1 and in Sq. NE4, south of Locus 2/3, and in the pit, Locus 11. The ceramic types included creamware, earthenware, shell-edged pearlware, painted porcelain, enameled, salt glazed, and Delftware.

Of the ceramic artifacts recovered, the most abundant type was creamware. In the 1750s, Thomas Whieldon of Williamsburg, Virginia, and Josiah Wedgewood refined creamware, producing "an even-firing [and] rich green glaze."[114] By 1762 Wedgewood had gone into business for himself in Burslem and developed creamware, also known as Queen's Ware and Leeds Ware. Creamware became one of the most regularly used ceramics for tableware in the late 18th and early 19th centuries in

---

[113] Watkins 1950:52; Hume 1969:102; Archaeology Lab Ceramics Database.

[114] Hume 1969:124; Richardson 2013; Diagnostic Artifacts in Maryland, http://www.jefpat.org/diagnostic/ColonialCeramics/Colonial%20Ware%20Descriptions/Creamware.html.

England and America. Creamware is commonly found at most early American sites.[115] It is no surprise, then, that creamware was discovered at the blockhouse site, establishing its use in the colonial American historical period and culture.

Of the 45 fragments recovered from the Greensboro blockhouse site, 23 (51.1%) of them may be classified as undecorated creamware.[116] The majority of these fragments appear to represent cups, bowls, and plates (Figure 5.46).

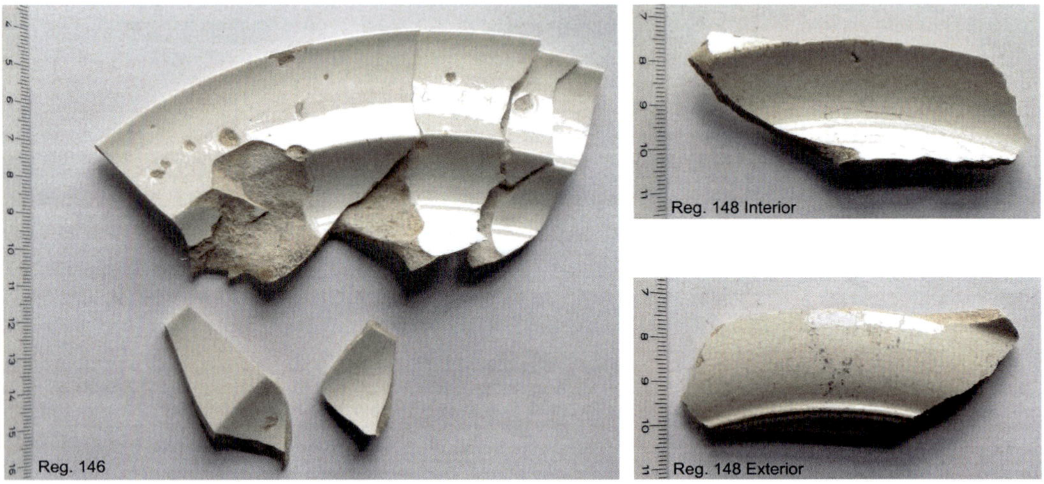

Figure 5.46. Two examples of creamware found at the blockhouse site. *Left*: Reg. 146, a shallow bowl or plate with horizontal flaring rim. *Right*: Reg. 148, the base and body fragment of a shallow carinated bowl. Photographs by Jill L. Baker.

These fragments appear to be of well-levigated clay, thus producing relatively thin walls and thus refined earthenware. The ware or paste type is buff/cream or white in color. The surface treatment consists of interior and exterior white glaze.

The next largest ceramic group may be described as decorated earthenware or creamware (Figure 5.47).[117] The examples were mostly rim, base, and body fragments of cups, small bowls, and plates or saucers. The ware consisted of a white, buff, or cream-colored clay/paste of well-

---

[115] Hume 1969:124–125; Richardson 2013.
[116] For all creamware, see Catalog of Artifacts, Regs. 5, 29, 51, 54, 96, 97, 98, 99, 117, 120, 146, 147, 148, 150, 151, 153 A–C, 167, 171, 172, 175. Contact archaeologicalhorizons@gmail.com.
[117] For decorated creamware, see Catalog of Artifacts, Regs. 30, 53, 53a, 57, 121, 132, 149, 152, 168, 169, 176. Contact archaeologicalhorizons@gmail.com.

levigated clay. Surface treatment consisted of a white background, glazed, with an over-glazed or under-glazed monochrome or polychrome painted decoration. Of the 45 ceramic fragments recovered from the site, 11 (24.44%) were decorated.

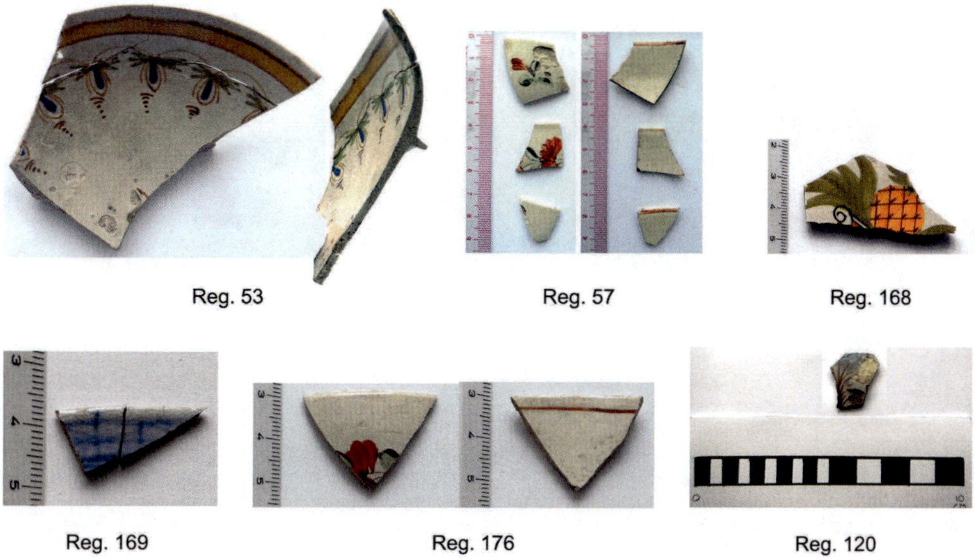

Figure 5.47. Select painted ceramic fragments found in the 2011, 2012, and 2015 seasons.

The next sizeable group was earthenware (Figure 5.48).[118] These examples were mostly rim and body fragments and only one handle fragment. The ware was coarse but well levigated and red and brownish-red in color. The rims' types were upright, thickened, and out-rolled (Regs. 56, 94), which probably represents jugs and pots, perhaps cooking pots. The fragments were heavily glazed on both the interior and exterior. The glaze varied in color from clear to dark reddish-brown to black. Of the 45 ceramic fragments, 9 (20%) were coarse, glazed earthenware.

[118] For earthenware see Catalog of Artifacts, Regs. 52, 56, 61, 89, 92, 93, 94, 95, 170. Contact archaeologicalhorizons@gmail.com.

Figure 5.48. Select examples of earthenware recovered from the blockhouse site during the 2011 and 2015 seasons. The ware is coarse but well-levigated clay with heavy interior and exterior glazing.

These fragments were recovered from the area that would have been between the two joists uncovered during the 2011 and 2015 excavation seasons. Based on the rim and fabric, it is possible that Registration numbers 56 and 95 are part of the same vessel; however, there are no physical joins.

The last ceramic type recovered during excavation is shell-edged pearlware (Figure 5.49). (Creamware initially boasted a shell-edge, which later became the most popular edging on pearlware.) Pearlware originated in England and was used from ca. 1780 to 1840. Early examples of pearlware (ca. 1780–1795) were painted, and the brushwork was drawn in toward the center of the vessel to create a wispy, feathery edge. However, as pearlware became popular, to save time, brush strokes were executed laterally rather than inward.[119] The shell edging was embossed or incised into the clay and adorned the rim edge or carination of bowls, plates, and saucers. The most commonly used colors were blue and green with red used only occasionally. Pearlware took many forms — mugs, jugs, bowls, plates — and was one of the most commonly used tablewares between 1790–1860. Pearlware is one of the most commonly found ceramic types at late 18th and early 19th-century sites.[120] Of the 45 ceramic

---

[119] Hume 1969:130–140; Richardson 2013. Jefferson Patterson, 2003,http://www.jefpat.org/diagnostic/postcolonial%20ceramics/shell%20edged%20wares/Shell%20Edged%20Wares%20Main.htm.

[120] Hume 1969:129–131; Jefferson Patterson 2003; Richardson 2013.

fragments found at the Greensboro blockhouse site, 2 were shell-edged pearlware.

Reg. 55 Reg. 139

Figure 5.49. Shell-edged pearlware. Fragments discovered during the 2011 and 2015 excavation seasons. Graphic by Jill L. Baker.

The two fragments shown in Figure 5.49 were discovered in the 2011 and 2015 excavation seasons located in the area between the two fieldstone floor joists. These rim fragments are from a plate and/or a bowl that would have had a horizontal flared rim. The fragments consist of a white glazed background, and the edges are scalloped, incised, and painted, one with blue and the other with green. The blue fragment, Reg. 139, appears to exhibit the inward-feathering decoration common to the early wares, and Reg. 55 appears to exhibit the horizontal brush stroke common to later wares.

This type of shell-edged pearlware may be further classified as Rococo Edgeware dating from 1775 to 1810. These may be compared to fragments found at the Bull's Head Tavern (18BC139) brick privy (Feature 9) (ca. 1800–1830) and Ruth Saloon (18BC79) brick-lined privy (Feature R-5) (ca. 1829–1837)[121] and a 19th-century rubbish pit in Middlebury, Connecticut, associated with a property owned by the Southmayd family until 1825, a

[121] Jefferson Patterson, 2003,http://www.jefpat.org/diagnostic/postcolonial%20ceramics/shell%20edged%20wares/Shell%20Edged%20Wares%20Main.htm.

prominent merchant family, or the Vansand family, a local manufacturer.[122]

Nails

Exactly when humankind began to use nails is not known; however, nails were used in ancient Egypt, Mesopotamia (Modern Iraq and part of Iran), ancient Israel, and of course in the Roman Empire. In ancient Egypt, nails were initially used for decorative purposes, to attach inlay, such as ivory, to furniture or boxes, or to secure gold or silver leaf to an item. These early nails were made from gold, silver, copper, or bronze, and date as early as ca. 3400 BCE.[123] Once iron came into use, spikes and nails were used more frequently to fasten structures. Reference to nails from Canaan and ancient Israel can be found in the Old and New Testaments of the Bible.[124] The Romans, however, used nails extensively throughout the empire. At Inchtuthil, a Roman fortress in Scotland dating to 83–86/87 CE, Sir Ian Richmond discovered a hoard of nails in a very well built and concealed pit.[125] As the Roman troops prepared to leave the fort, they buried some 875,000 iron nails (10 tons) so that the Caledonians would not move in and melt down the iron to make weapons.[126] These references reflect the use of nails from the Iron Age (ca. 1200–586 BCE) onward. It was the Romans, however, who used nails extensively throughout their empire. English settlers to the American colonies brought from England as many nails as possible in the early 1600s.[127] Supplies continued to come from England until the time of the Revolution, when local supplies became dependable. One such nail supplier was John Little of Philadelphia in the 1770s.[128]

Iron nails were forged by hand, also known as hand-wrought, from antiquity until ca. 1800/1815 when nails were produced by a cutting process. The size, shape, and manufacturing technique did not change much from antiquity until the 19th century. The hand-made wrought iron nail process began by heating iron ore with carbon, causing it to become wrought, and shaping it into round or square rods or bars, approximately the width of the nail's shaft. In 1580 the slitting mill was invented in present day Belgium, and one began to operate in Dartford, England, in

---

[122] Archaeology and Anthropology Collections. Saint Mary's University, Department of Anthropology, http://www.smu.ca/academics/departments/anthropology-shell-edged-pearlware.html.
[123] Ogden 2006:164; Gale, et al. 2006:325, 356, 367.
[124] Judges 4:21–22, 5:26, 16:13, 14, 2 Chronicles 3:9, Isaiah 22:23, 25, Jeremiah 10:4, 17:1, Mark 15:13–14, 29, John 19:10, 20:25.
[125] CastlesFortsBattles.co.uk. LeFever 2008.
[126] LeFever 2008.
[127] LeFever 2008.
[128] Nelson 1963.

1590. These machines used waterpower and a large cleaver to slit large sheets of iron into rods. The rods were then sold to nailers or nailsmiths. The rod was about five feet long and one quarter inch square or round. The nailer put one end of the rod into the fire until heated. It was then removed and hammered, or drawn, to the desired shape and length. Then the rod was placed into a header, which held the rod in place more easily, and the rod was then set on top of a chisel attachment on an anvil called a hardy. The rod was then hammered on top of the chisel, severing the newly formed nail shaft from the rod. With one quarter of the nail shaft above the header, the nail was inserted into a hole in the anvil and the nail head formed by pounding the header. The variety of size and shape into which nails were made were task-specific. Smaller, finer nails were used as finishing nails for wood trim, floorboards, or furniture. Larger nails were used for structural fastenings and roofing.[129]

Hand-forged nails and the machine-cut nails of the 1800s were stronger than cut or wire nails, a process that came later, and were far superior to modern nails. Hand-forged nails and the early machine-cut nails were formed most commonly into a rose-head shape, a four-sided pyramidal shape, or into 'butterfly', L-heads, or T-heads. The rose-head shape had four straight edges that were relatively sharp. When pounded into the wood according to the grain, the sharp edges of the nail sliced through the fiber of the wood, thus creating a very strong bond when the wood swelled. Removing a square nail that had been pounded into wood was not at all easy. This is one reason it was easier to burn down a building to retrieve nails than it was to pull them out. In the 17th century, the Kent County courthouse in Delaware was ordered destroyed so that the nails could be collected and reused,[130] and the Virginia House of Burgesses outlawed the burning of vacant plantations for the purpose of retrieving nails, and instead offered to provide former owners with a quantity of nails equivalent to that which had been used to construct their buildings.[131]

As with most colonial American sites, an abundance of nails were recovered during excavation. As with pottery, nails have been chronologically and typologically identified and categorized. The nails found at the Greensboro blockhouse fit into the chronological and typological schemes described by Hume (1969), Nelson (1963), and Wells (1998).

The nails recovered from the Greensboro site (Figure 5.50) fit into five basic categories based on the shape of their heads: rose, L-shape, T-shape, no head, and round. Of the 95 nails discovered thus far, 18 (18.94%) were rose shaped; 31 (32.63%) were L-shaped; 21 (22.10%) were T-shaped; 11

---

129 LeFever 2008; Wells 1998; Hume 1969.
130 Peterson 1950:23.
131 LeFever 2008:60.

(11.57%) had no head; 5 (5.26%) were round; and 9 (9.47%) were too badly damaged to determine.[132] The nails ranged in size from 3.3 cm to 9.4 cm. The nails recovered in the 2011 and 2015 seasons seemed to occur in clusters throughout the sub-floor fill, usually in proximity to the stone joists. The nails comprising the clusters were generally a combination of small, medium, and large sizes. Examples: Registration numbers 26, 27, 28; Registration numbers 32, 33, 34, 35, 36; Registration numbers 129, 130, 131; and Registration numbers 133, 134, 135, 136, 137 were found in clusters in different parts of the site but in association with one of the two joists. The nails in these clusters included large L-headed nails or nails with no heads, and small and medium rose-head nails with tapered pointed ends. Presumably the presence of the different sizes and shapes of the nails suggests the sort of task they were meant to perform, such as structural versus securing a floorboard.

The nails recovered from the site were discovered mainly in the 2011 probe and Square NE4, excavated in 2015. Most of the nails were located in the area between the stone joist (Locus 1) of the 2011 probe and Locus 2, stone joist, in Sq. NE4. Theoretically, if Locus 2 was an outer wall, it would have supported the first floor wall and functioned as a sill for the wood flooring. The joist uncovered in Sq. NE4 also would have functioned as a sill for the wood flooring.. The wooden flooring, then, would have spanned the two joists. If at the time of collapse (ca. 1815) this timber was removed and repurposed elsewhere, the nails were removed and dropped onto the sub-floor fill. These discarded nails remained *in situ* until excavated in 2011 and 2015.

The salvaged timber may be evidenced by the condition of the nails. In total, 95 nails were recovered. Of these, 78 (82.105%) were bent and/or broken. Considering how difficult it was to remove square nails from timber, especially if they were clinched, it is likely the nails were bent and/or broken during their removal. Assuming it was the timber which was desired for reuse elsewhere rather than the nails, the building would not have been burned to the ground. Instead, nails were removed to obtain the wood.

As mentioned above, thus far, no wood has been discovered during excavation; however, plenty of nails, most of which were disfigured, have been. There is another possible explanation for some of the nails being bent. To ensure that nails did not come loose, the carpenter used nails that were longer than the boards they secured, thus causing the tip to protrude from the other side of the wood. The tip was hammered up until it was flush with the wood or clinched over (also called a clinched nail), thus preventing the nail from wiggling loose. This was common practice

[132] All of the nails recovered from the Greensboro blockhouse excavation can be found in the Catalog of Artifacts. For more information contact AHI at archaeologicalhorizons@gmail.com.

on gates and doors. This was called 'dead nail', from which the phrase 'dead as a doornail' is derived. Additionally, by measuring the distance between a nail's head and the clenched tip, one could estimate the width of the board(s) it secured. It is likely that both scenarios, pulling nails out of the wood and clenching, may account for the high percentage of bent nails found at the Greensboro blockhouse site.

Most of the nails appear to have been hand-forged wrought iron nails (Figure 5.50). These probably date to the late 18th century prior to the use of machine-cut nails around 1810. Several of the nails, however, may have been early machine-cut nails. These include Registration numbers 32, 104, 109, for example.

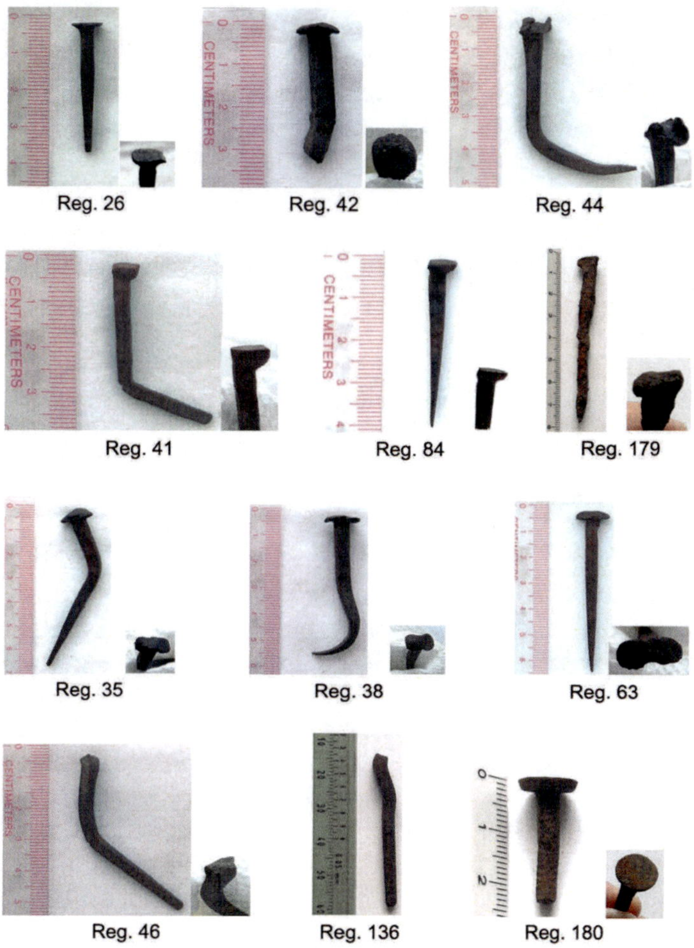

Figure 5.50. Selected nails. *Top row*: rose-head nails, Regs. 26, 42, 44. *Second row*: L-head nails, Regs. 41, 84, 179. *Third row*: T-head nails, Regs. 35, 38, 63. *Fourth row*: nails without heads, Regs. 46, 136. Reg. 180, round-headed nail. Graphic by Jill L. Baker.

Artifacts – Miscellaneous

In addition to the pottery and nails, several other artifacts deserve to be discussed. These include a whisk, two hoe heads, a pair of scissors, an unidentified metal fragment, bricks, and stone tools. These provide information regarding the type of work being conducted at the site as well as locally sourced materials and those that were likely brought from a distance.

Metal Artifacts

During the surface survey in 2010, in the area that became Sq. NE4, the metal head of a whisk (Reg. 10) was recovered (Figure 5.51). A thin, coiled metal wire formed the whisk. The coil was narrower at one end and wider at the other. At the narrow end, a screw or tack with a scored head was attached to the coil. Perhaps this fastened the coil to a wooden or metal handle, which did not survive. The dimensions were 27/28 mm at the narrow end and ca. 36 mm at the wide end. Early whisks were made of bundled twigs, often from apple or peach trees. The ends of the twigs were clipped and mashed to release the flavor of the sap into the batter being whisked. In the 19th century, the wire whisk was introduced to early American kitchens.

According to Hume, hoe blades have not received proper study, which appears to remain the case in 2015. However, he does mention that a typical 18th-century hoe blade measured 8 inches by 12 inches and was "divided at the top, and capped by flat iron shoulders that snuggled well into the instep of the foot."[133] Hoes were generally used to break ground, loosen the earth, and remove weeds. The word hoe comes from the German verb *hacken* (to hack, chop, peck at), and the noun *Hacke* refers to a hoe and/or a pickaxe.[134] Hoes have been in use since antiquity, at least as early as the 5th millennium BCE. Originally, they were made from stone, bone, and horn. By the 14th century BCE, hoes were made from bronze, and later, from iron in the 10th century BCE. Hoes were utilized in Mesopotamia, Egypt, Canaan, and Europe, anywhere agriculture was practiced. In Mesopotamia, Sumerian mythology credits the invention of the hoe to Enlil, chief among the gods. The Babylonian king Hammurabi created a set of law codes ca. 1750 BCE, in which hoe tools are mentioned. In Pre-Dynastic Egypt, hoes are depicted in innumerable farming scenes. Finally, hoes are mentioned in the biblical book of Isaiah dating to approximately the 8th century BCE. Over time, the hoe took on numerous shapes and sizes from square to half circles with straight blades to triangular shapes with V-shape blades. The hoe has been in

---

[133] Hume 1969:275.

[134] Laws 2014:63.

continual use throughout France, Germany, and England, and its technology was imported to the Americas.

In the 2011 season a pair of scissors was discovered in the western balk at the 10.8-meter marker from the northwest corner grid stake. This pair of scissors is corroded and remains unconserved. The pair measures 11.1 cm (4.33 inches) in length (Figure 5.51).

Reg. 6

Reg. 144

Reg. 10

Reg. 45

Figure 5.51. Miscellaneous artifacts. *Top row*: hoe blades. *Bottom row*: partial whisk and scissors. Photographs by Jill L. Baker.

Scissors developed from shears, which have been in use since antiquity. The earliest known shears were from Egypt and Mesopotamia, and date to ca. 1500 BCE. Shears were made from a single, thin, flexible band of bronze that was bent in half. The end of the strip (the blades) aligned, and the bent end of the band acted as a spring. The blades were squeezed to create the cutting action.

Scholars suggest that the Romans introduced scissors in the first or early second century CE, and their use quickly spread throughout the Roman Empire. Shears were mainly useful for shearing sheep's wool. Barbers also used shears to cut hair. Shears remained in use throughout Europe as late as the 17th century. However, scissors, whose blades were fixed at a pivot point, eventually replaced shears, probably because they were easier to handle.

Scissors were transported to the colonies by early European settlers and are found at numerous colonial and post-colonial sites.[135] These scissors were also made of forged iron and took many sizes and shapes. By 1760, William Whiteley & Sons (Sheffield) Ltd. were manufacturing scissors and distributing them throughout Europe and America. By the mid-17th century, the blades were wider and thicker, and the rivet was located just below the handle branches.[136] The blades, or hafts, were rectangular in section, and the handles were made from a separate piece of iron and offset to the blades' axis.[137] Scissors were used for sewing, in the kitchen, and for personal hygiene, and were a general all-purpose tool around the homestead. Scissors have been found at numerous colonial sites including the Mount Vernon Midden.[138]

In the 2015 season an unidentified metal object was discovered in the northwest corner of Sq. NE4. The metal fragment was lodged among and under Locus 1 and in Locus 11 in the northwestern balk corner. The fragment measures approximately 15.5 cm long by 7 cm wide. It is concave and appears to have a carination[139] at one end. There appear to be faint burn marks on the exterior at the mid-section of the body and at one end (Figure 5.52).

Presumably this was a metal vessel, perhaps a cooking pot or jug that was left behind when the structure was dismantled. It was discovered in the area assumed as that of the hearth and associated with the ash/charcoal pit.

---

[135] Hume 1969:267–269.
[136] Hume 1969:167.
[137] Hume 1969:267.
[138] mountvernonmidden.org.
[139] A carination in pottery or glassware is a ridge or angle that changes the direction of the vessel wall.

Figure 5.52. Unidentified metal fragment. Faint burn marks appear on the vessel's exterior at the body's mid-section and upper tip. *Left*: exterior. *Right*: interior. Graphic by Jill L. Baker.

Bricks

Bricks and brick fragments were found throughout the site, the majority of them located in Sq. NE4, the area presumed to be the hearth/chimney, and the area between the two floor joists (Figure 5.53). The bricks were whole and partial. The whole bricks measured ca. 18–20 cm long (7.14 to 7.87 inches) by 9 to 10 cm wide (3.54 to 3.93 inches) by 14 to 14.5 cm thick (5.5 to 5.7 inches). Although varied in size and shape, the partial bricks measured approximately 9 cm by 10 cm by 4 to 5 cm (3.54 to 3.9 inches by 3.5 to 3.9 inches by 1.57 to 1.96 inches) and were roughly square, whereas the whole bricks were rectangular. These bricks appear to have been made by hand using a mold, and some may have been cut with a wire or string. The color of the clay is red to brownish-red, and the bricks appear to have been fired. Some bricks show large inclusions (Figure 5.53). None of the bricks show evidence of mortar. At present, the function of the bricks from the Greensboro site is not clear. The majority of the bricks were found among and around the hearth/chimney stack. Perhaps the chimney was made from bricks; perhaps the fireplace was lined with brick; or perhaps the post-war inhabitants modified the fireplace with a brick lining or by adding a brick oven adjacent to the main firebox. It is also not certain how many bricks were removed when the structure was harvested for materials. Only further excavation will be able to answer these questions.

Figure 5.53. Selected bricks from the 2010 surface survey and 2015 excavation season. Bricks 1 and 2 were whole, but most of the bricks found during the 2015 season were partial. Graphic by Jill L. Baker.

Bricks were among the first human-made construction materials. As early as the Pre-Dynastic period in Egypt, humans were making bricks from the mud found near the Nile River to line tombs and build walls. By the First Dynasty in Egypt, mud bricks were manufactured and used to construct architecture (Figure 5.54). These early bricks measured ca. 50–65 x 9–10 x 10–15 cm thick and were held together with a mud mortar. Mud bricks were also used in Mesopotamia, and were fashioned with five flat sides and one convex in a conical shape, and rectangular. Mud bricks were made by hand. The clay/mud was gathered, washed to remove as many inclusions as possible, and strengthened with a binder such as lime or reeds. The prepared clay/mud was pressed into a wood mold, removed, and set out to dry in the sun's scorching heat.

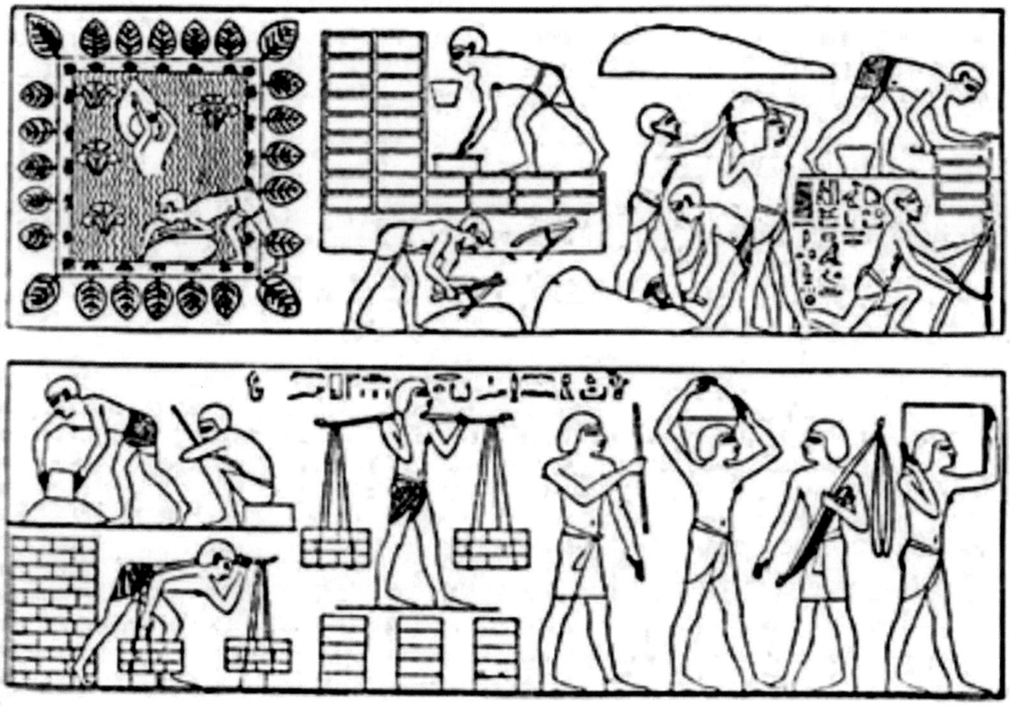

Figure 5.54. Brick-making scene from the tomb of Rekhmira (TT100; Davies 135:Pls. XVI, XVII, XXIII), a nobleman from the mid-18th Dynasty (ca. 1479–1425 BCE). This scene depicts mud/clay being harvested from the Nile, cleaned, pressed into wooden brick molds, set out to dry, and placed into architectural walls. Graphic adapted from http://www.bible-people.info/moses_plagues.htm.

Throughout the history of the ancient Near and Middle East, mudbrick was one of the primary building materials. Entire buildings, walls, and gates were constructed with mudbrick. However, mudbrick was susceptible to the weather; rain, water, and even sand could cause great damage to a mudbrick structure. To combat the elements, a thick lime plaster was added to the exterior.

Over time the size, shape, and manufacture of bricks changed. By the Roman period, bricks had shrunk in size and were baked in kilns rather than by the sun, which made them stronger and resistant to the elements. The use of brick in Rome increased during the time of Nero, ca. 64 CE, when much of Rome was destroyed by fire. Rome was rebuilt with brick, which was more fire resistant.[140] The use of bricks spread throughout the Roman Empire, and the size and manufacture of bricks changed little. Knowledge of bricks and their manufacture came with the English, French, and Dutch. Brick sizes and shapes varied slightly, but in general, bricks found in mid-18th century Virginia measured 71/2 x 31/2 x 2 inches, as well as those sometimes identified as English, and Dutch/Flemish bricks measured ca. 71/8 x 31/4 x 13/8 inches.[141] Similar bricks have been discovered at colonial sites and at the Mount Vernon Midden; for example, one 18th-century brick measured ca. 7.55 x 3.50 x 1.9 inches, and a partial brick measured ca. 6.37 x 4.133 x 2.16 inches, similar to the size and shape found at the Greensboro site.[142]

Stone Tools

Among the myriad of rocks presumed to have comprised the walls and joists of the blockhouse structure, several stood out as tools or architectural components. Given that the area was first inhabited by native peoples, it would not be surprising to find tools that were either left behind or lost. Additionally, based on a statement made by Jonathan Elkins in his memoirs, General Bailey wrote to General Washington suggesting and receiving approval for the Native Americans to be treated kindly and respectfully. This led to a friendship with some local tribes, including the Cagnawagah.[143] This friendship included trading of goods and ideas, which helped the early settlers to survive. It is likely that goods traded may have included stone tools or at least knowledge of how to make stone tools. The stone tools found at the blockhouse site may have been made and used by prehistoric peoples who inhabited or visited the site or by the blockhouse inhabitants themselves.

---

[140] Oleson et al. 2008:264; Tacitus, *Ann.* 15.42.
[141] Hume 1969:82–83.
[142] Mount Vernon Midden Project, www.mountvernonmidden.org, Object ID 2767 and 2768.
[143] Elkins 1921:189–190.

During the 2012 season two stone tools were recovered from the probe (Figure 5.55). These were found in Locus 5, which may have been a dwelling surface. One stone tool, Registration 122, was oblong, roughly rectangular, and smoothed with a flat base and rounded upper. Tools such as this, used for grinding or smoothing, fit perfectly into one's hand.[144] Striations on the flat side of this stone extending the length of the surface suggest that it may have been used for smoothing or grinding. It measured ca. 12 cm long x 5 cm wide at the narrow flat side and 6 cm wide at the widest point on the flat side. It was between 2.7 and 3.3 cm thick. The second stone tool, Registration 123, found during the 2012 season, was oblong with one wide end that tapered to a narrow end (Figure 5.55). It measured 8.4 cm long, 2.6 cm at the narrow end, 4.4 cm wide at the widest end, 1.4 cm thick at the narrow end, and 2.3 cm thick at the wider end. On the narrow end, smoothing and incised marks suggest that a handle of wood, bone, or antler may have been attached to it. Perhaps this was used as some sort of chopping or pounding tool, or was possibly a celt-form axe.[145]

Figure 5.55. Stone tools. *Left*: Reg. 122. *Right*: Reg. 123. Graphic by Jill L. Baker.

In the 2015 excavation season, a stone tool was discovered in Sq. NE4 (Figure 5.56). This stone is flat on one side and convex on the top with wide, rounded sides. It fits comfortably in the grip of one's hand. This stone was found among Loci 2/3 and appears to have been a rubbing stone or grinder. It measures ca. 7.6 cm in length and ca. 5.0 cm at its widest point.

---

[144] cf. Ground Stone Artifacts, https://archaeology.uiowa.edu/ground-stone-artifacts-0.
[145] cf. Celt Forms, http://www.peachstatearchaeologicalsociety.org/index.php/9-hardstone/144-celts.

Reg. 204

Figure 5.56. Stone tool. Reg. 204. *Left*: top of stone. *Middle*: flat bottom of stone. *Right*: side of stone showing flat bottom and curved top. Graphic by Jill L. Baker.

Finally, throughout the site we found small to medium-sized fragments of quartzite stone. At first these seemed insignificant; however, on further reflection and research, it is likely that these represent a core and flakes, likely used to make stone tools. The largest concentration of these was found in the area between the joists and in Sq. NE4, especially during the 2011 and 2015 seasons.

When stone tools are made, especially blades and projectile points, the flakes represent the debitage produced by refining the actual blade or projectile points. Given that these flakes and perhaps a core were found securely within habitation levels of the structure may be further evidence that the inhabitants were making some of their tools from stone (Figure 5.57).

Similar flakes have been found at the Mount Vernon Midden. However, those flakes have been attributed to prehistoric peoples who inhabited that site.[146]

[146] Mount Vernon Midden Project, www.mountvernonmidden.org Object ID 3209.

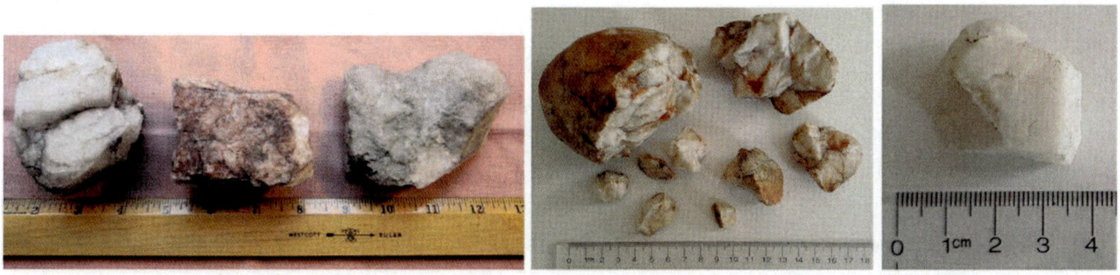

Figure 5.57. Quartz core and flakes. From the 2011, 2012, and 2015 excavation seasons. *Left*: three quartz flakes from the 2011 probe. Photograph by S. Haslam. *Middle*: core, broken with flakes from the 2012 season. *Right*: quartz flake from the 2015 season, Square NE4. Middle and right photographs by Jill L. Baker.

As suggested above, the inhabitants of the blockhouse may have been making stone tools, a craft that may have been learned from the local Native American population who befriended some of the soldiers and early settlers. Jonathan Elkins (1921) described how a benevolent relationship formed, and based on some of the letters between General Bayley, Colonel Hazen, and General Washington, it would not be outside the realm of possibility that poorly supplied troops employed local technology, specifically in the form of stone tools.

Also discovered during the 2015 season was a large stone that contained what appeared to be intentional incised pictographs or post marks, or accidental incised marks due to wear and tear (Figure 5.58). This sedimentary stone absorbed moisture and looked permanently wet when in the field. Only when taken inside did it dry completely. The stone was irregular in shape with one side more pointed and the other rounded. It measured ca. 62.5 cm long and 29.5 cm wide at its maximum points. The top appeared to be flat while the sides and bottom remained rough.

The top, flattish surface of the rock appeared to contain incised marks. When looking at the stone with the pointed side to the left, on the right side of the surface there appears to be an incised irregular circle, and on the left an incised oval or rectangular shape with two or four small holes in the center of the incised area. Extending down from the left side of the incised rectangle/oval is an incised dashed line. On the bottom edge of the rock there are several incised, chevron-like markings, which almost appear to have been made by a stamp or possibly from a tool hitting the stone with some force. This chevron pattern also appears on the rounded edge of the stone.

Reg. 203

Figure 5.58. Rock with incised markings. *Upper left*: the stone *in situ*. *Upper right*: top surface showing the incised chevron patterns, circle, and oval shapes. *Lower left*: the rough unworked side of the rock. *Lower right*: rock *in situ* showing the Y-shaped incision on the front side of the stone. Graphic by Jill L. Baker.

This stone was in use as part of Locus 2 in Sq. NE4. It sat on the edge of the joist and may have been structural in function. It may have served as a foundation stone and/or for posts that supported the second floor and/or floor beams. It may also have been used elsewhere before being placed here for secondary use. Given that it was one of the upper stones in the joist, it is likely it was used as a base for posts that supported the second floor. The incised marks may represent nail and hammer marks, or some other tool with a fixed pattern, while the incised round and oval/rectangular marks may represent negative impressions of the actual posts that have long since been removed. Additionally, on the lower front side of the stones there appears to be an incised elongated Y, possibly a mason's mark.

Chapter 6

Comparative Analysis of Contemporaneous Blockhouses

As with the artifacts, it is important to set the architectural layout and features of the Greensboro blockhouse into a wider contemporaneous context. As discussed earlier in the book, fortifications occurred in three basic groups: garrison houses, blockhouses, and fortified or military forts that included numerous buildings, earthen works, and stockade fences. The architectural features of these structures were task specific and can be set into a chronological and function-specific framework. Based on comparative analysis, the architectural layout and features of the Greensboro blockhouse discovered thus far are consistent with those of other blockhouses that have either remained intact or been restored. The discussion in this chapter will focus on architectural and archaeological features specific to blockhouses, and will focus on stand-alone blockhouses and those incorporated into larger complexes.

Blockhouses, especially stand-alone blockhouses, were meant to literally block the passage of opposing forces such as military troops or even rival trappers. Blockhouses were used to accommodate military personnel and their paraphernalia, to provide defensive protection in general, or to provide shelter.[147] When explorers, traders, and religious purists populated the eastern shores of North America, they settled and spread into new regions, often encountering aboriginal inhabitants and fellow colonists who also wished to exploit the rich resources of this unspoiled continent and lay claim to fertile land. Soon after their arrival, the colonists' relationship with the original inhabitants and fellow pioneers soured, and

---

[147] Chartrand 2011.

ruling entities from Europe sought to maintain control, which demanded a need for solid, fortified structures for protection from rival or warring opponents. To satisfy this need, blockhouses were constructed by the English, French, Spanish, and Canadians at key defensive points along the eastern coastline of the North American continent and along important river passages. Some of these include Fort Halifax in Maine, Fort Edward in Nova Scotia, Royal Blockhouse near Moreau, New York, and Fort Pitt in Pennsylvania. Others were erected inland along important military roads, such as Fort at No. 4, New Hampshire, or to protect a community, such as Bridgeman's Fort in Vernon, Vermont, which was meant to protect the townspeople from raids by native peoples.[148] For the purpose of this discussion, the best contemporaneous examples of blockhouses include those located at Fort Pitt, Fort Kent, Fort Halifax, Clergue, and the LaColle River Blockhouse.

The Fort Pitt Blockhouse

The Fort Pitt blockhouse was built sometime between the years 1763 and 1764 on a spit of land between the Allegheny River and the Monongahela River (Figure 6.59). The structure was pentagonal in shape; the back and two side walls formed right angles that measured 16 feet in length, and the two front walls formed a triangle that measured 15 feet in length.[149] The overall structure measured 23 feet wide and 26 feet deep, and the base measured 483 square feet.[150] The foundation was constructed of worked and unworked limestone and a limestone mortar. The foundation walls were approximately 2 feet thick and incorporated an inner and outer layer joined by header stones, called a casemate wall. A casemate wall looks like a ladder, and the chambers between the rungs, or header stones, were filled in with a mixture of smaller stones, lime, and sandy soil. Casemate walls were in use in Israel as early as the Iron Age (ca. 10th century BCE to 1st century CE), and can still be seen at sites such as Masada and have recently been excavated at Khirbet Qeiyafa.[151]

In 1764, the original foundation of the Fort Pitt Blockhouse reached more than 7 feet above ground level on the exterior. However, over time silt built up around the structure, and soil was deliberately deposited to resist flooding, thus raising the foundation as it is today to approximately four feet above ground level. On top of the stone foundation walls was a wooden course hewn from white oak that contained gun loops, and each gun-loop opening was individually cut. Atop the wooden gun loop course was another course of brick forming the second story, on top of which

---

[148] Thompson 1842:67–68.
[149] Weaver 2013:28.
[150] Weaver 2013:28.
[151] Khirbet Qeiyafa Archaeological Project, http://qeiyafa.huji.ac.il/.

was another wooden gun loop. On top of that was another course of brick into which the wooden rafters were incorporated which supported the roof.[152] The first floor was made of wooden beams and wooden planks set on top of them. The gun loops set into the wood course on top of the stone foundation measured 7 feet from floor level. To reach the gun loops, soldiers stood on top of a firing step. The interior floor was built up so that the walls were only 5 feet, 4 inches in height, and floor level is approximately 16 inches below ground level.[153] The doorway was situated under one of the gun loops, which appears to have served as a sort of lintel for the entrance, although the doorway was relocated later. Nevertheless, the original position of the doorway would have been well defended under the gun loop.[154]

During archaeological excavation, their team discovered the remains of a fireplace that dated to the middle 18th century. Later, in 1894, the fireplace was removed and replaced by steps, which provided access to the second floor for tourists. The steps were located in the same place as the fireplace remains, suggesting that the fireplace was located on a wall opposite the doorway.[155]

From the 1780s onward, the blockhouse served as a residence to numerous tenants, the earliest being William Turnbull in 1785 followed by Isaac Craig and his family sometime around 1786.[156] The Craigs constructed a brick house next to the blockhouse, which was larger and probably better suited for the needs of a growing family; however, by 1789 the Craig family moved out of the house complex.[157]

In 1892 the blockhouse was presented as a gift to the Daughters of the American Revolution (DAR) of Allegheny County, Pennsylvania, to preserve it for future generations. The DAR carefully removed the additions and began to restore and preserve the blockhouse. Archaeological excavations have revealed important information regarding the construction of the blockhouse and have yielded numerous artifacts. Most of the artifacts recovered from excavation pertain to the civilian families who made the blockhouse their home. Some of these artifacts include pottery fragments, glass bottles, a bone tool possibly used to tan hides, a bear tooth that may have been worn as a pendant, and the remnants of four clay tobacco pipes with the initials TD inscribed on one of the fragments. From the period of military occupation of the blockhouse, approximately twelve items were discovered including shot

---

[152] Weaver 2013:28–30.
[153] Weaver 2013:30
[154] Weaver 2013:30
[155] Fort Wiki, http://www.fortpittblockhouse.com/archeology/.
[156] Weaver 2013:40.
[157] Weaver 2013:40.

and gunflints.[158] Further excavations revealed the original wood flooring, the fireplace added during the residential phase in the middle to late 18th century, and a brick floor that covered the original timber floor which was added in the middle 1800s during the residential phase.[159]

Figure 6.59. Fort Pitt Blockhouse as it appeared in 1903 after the DAR restoration. Adapted from: *Left*: "FortPittBlockHousecirca1903" by J. C. Bragdon - Crop of page 74 from Views of Pittsburg published in 1903, full text available at Historic Pittsburgh digital collection at the University of Pittsburgh's digital library http://www.library.pitt.edu/libraries/drl/.. Licensed under Public Domain via Commons https://commons.wikimedia.org/wiki/File:FortPittBlockHousecirca1903.jpg#/media/File:FortPittBlockHousecirca1903.jpg.

Fort Kent, Maine

The Fort Kent Blockhouse is located at the convergence of the St. John River and Fish River in Fort Kent, Maine (Figure 6.60). Constructed by Captain Nye, who commanded a Maine Civil Posse that inhabited the

[158] Fort Pitt Blockhouse. http://www.fortpittblockhouse.com/archeology/.
[159] Fort Pitt Blockhouse. http://www.fortpittblockhouse.com/archeology/.

blockhouse in 1839, this is the only remaining structure from the larger Fort Kent outpost erected after a border dispute between the United States and Canada known as the Aroostook War. Nye and his posse were replaced in 1839 by a garrison commanded by Captain John Winder, who expanded the post to include barracks, officer's quarters, and other buildings. In 1841 the US Army took control of the blockhouse and expanded it, but by 1843, they had abandoned it. In 1845 it was decommissioned as a fortification, and in 1858, it was sold to private individuals.[160]

Figure 6.60. Fort Kent Blockhouse. *Left*: 1910 post card (adapted from post card for sale on Ebay http://www.ebay.com/itm/1910-Postcard-The-Block-House-Fort-Kent-Maine-ME-/230452043065). *Middle*: photograph from 1911. Adapted from Penobscot Maine Museum, Maine Memory Network (https://www.mainememory.net/artifact/31460). *Right*: Fort Kent Blockhouse present-day (adapted from ViewPhotos.org. Copyright belongs to J. D. Walters).

The remaining blockhouse measures 23.5 square feet at the base and consists of two stories, with the second story overhanging the walls of the first by 15 inches. The walls measure 19 inches and were made from square-hewn cedar timbers. These were placed on top of a shale rock foundation 1–1.5 feet in height to protect the lower timbers from decay. The base of the blockhouse contained a powder magazine.[161] The roof was pyramidal in shape, and each side contained a gabled dormer; however, the dormers were removed in 1926 when the roof was replaced. The main entrance faced west, and was flanked by 4 gun loops. The other walls contained 12 gun loops. The second floor contained doors that measured 24 x 30 inches, but these were eventually replaced with windows on the east and west sides. The east and west second floor walls each contained 11 gun loops, and the north and south walls each contained 15. The first-floor walls also contained cannon ports on the north and south sides, which measured 12 x 15 inches. (These details are from the National Registry of Historic Places.)

[160] Fort Wiki.
http://www.fortwiki.com/Fort_Kent; http://acim.umfk.maine.edu/blockhouse.html, http://www.nps.gov/maac/planyourvisit/blockhouse.htm.
[161] Fort Wiki. http://www.fortwiki.com/Fort_Kent;
http://acim.umfk.maine.edu/blockhouse.html,
http://www.nps.gov/maac/planyourvisit/blockhouse.htm.

Fort Halifax

On the order of Governor Shirley of Maine, Major General John Winslow and 500 to 600 men began construction of Fort Halifax in 1754, but the work was completed by Captain Lithgow, who replaced Winslow (Figure 6.61). It was located at the convergence of the Sebasticook and Kennebec Rivers. The site was formerly an ancient Native American encampment and village just south of a series of waterfalls and rapids, and was a well-known passage of the Norridgewock and Penobscot tribes as they journeyed to Quebec. The fort functioned as a trading post and provided protection to southerly settlements from the French and Indians who made incursions into the area during the French and Indian War (1754–1756). In September 1775, just before the start of the American Revolution, Col. Benedict Arnold and his expedition stopped at Fort Halifax on their way to Quebec City.[162]

The fort's final arrangement included a palisade (stockade fence) 120 square feet (800 feet long); a row of single-story barracks; a two-story fort-house that measured approximately 40 x 80 feet, which served as officers' quarters, storage, and the armory; a guard house at the entrance, and two two-story blockhouses in the northeast and southwest corners of the fort. In addition to being two–story and cantilevered with a pyramidal hip roof, the blockhouses were constructed of square-hewn timbers dovetailed at the corners, with small windows and gun loops. On the hill above the fort were two smaller redoubts. The first floor of the blockhouse measured 20 square feet, and the second level measured 27 square feet.[163]

Much of Fort Halifax was removed except for the southwest blockhouse, the one nearest the river. It remained standing and was visited by tourists, many of whom left their mark on the structure. The Daughters of the American Revolution owned the blockhouse from 1924 to 1966 and managed to keep it in good repair. In 1966 the DAR gave the property to the state of Maine, and in 1968 the blockhouse was declared a National Historic Landmark.

---

[162] Sprague 1921; Fort Wiki.
http://www.maine.gov/dacf/parks/discover_history_explore_nature/history/fort_halifax/index.shtml; http://fortwiki.com/Fort_Halifax_%281%29
[163] Sprague 1921; Fort Wiki.
http://www.maine.gov/dacf/parks/discover_history_explore_nature/history/fort_halifax/index.shtml; http://fortwiki.com/Fort_Halifax_%281%29

Figure 6.61. Fort Halifax. *Left*: Drawing of Fort Halifax as it appeared in 1755. The remaining blockhouse is the southwestern one in the corner. Adapted from Sprague 1921. *Right*: Fort Halifax as it appeared in 1936. Adapted from Wikipedia ("Fort Halifax, U.S. Route 201, Winslow (Kennebec County, Maine)" by Josiah T. Tubby, Photographer - http://memory.loc.gov/cgi-bin/displayPhoto.pl?path=/pnp/habshaer/me/me0000/me0062/photos&topImages=088235pr.jpg&topLinks=088235pv.jpg,088235pu.tif&title=2.%20%20Historic%20American%20Buildings%20Survey%20Josiah%20T.%20Tubby,%20Photographer%20October%2021,%201936%20VIEW%20OF%20BLOCKHOUSE%20(river%20side)%20%20%3cbr%3eHABS%20ME,6-WINLO,1-2&displayProfile=0. Licensed under Public Domain via Commons - https://commons.wikimedia.org/wiki/File:Fort_Halifax,_U.S._Route_201,_Winslow_(Kennebec_County,_Maine).jpg#/media/File:Fort_Halifax,_U.S._Route_201,_Winslow_(Kennebec_County,_Maine).jpg).

Clergue Blockhouse

Beginning in 1779, a group of independent fur traders formed the North West Company, a partnership aimed at strengthening their profession and industry (Figure 6.62). Sault Ste. Marie was a common stopping point for traders journeying from Montreal to northern and western points, so a partnership seemed sensible. The partnership dissolved and renewed several times until 1804, when the North West Company reformed under better management and survived. Until 1790, the North West Company maintained houses and stores at Sault Ste. Marie but moved to the British side of St. Mary's River after the Jay Treaty. There the Company constructed a sawmill, docks, storehouses, and dwellings, as well as a canal and lock, which controlled passage of bateaux through the rapids. The post was destroyed by American troops during the War of 1812, but the Company rebuilt and occupied the post until 1821 when the North West Company was purchased by the Hudson Bay Company. The post remained occupied until 1867 when the last of the employees left. Although a caretaker was hired to maintain the post, it quickly fell into disrepair, and the buildings either collapsed or were deliberately demolished except for the blockhouse, which had also served as the powder magazine.[164]

The property was purchased by Francis H. Clergue from the Hudson Bay Company in 1894. An employee of a group of financial investors from Philadelphia, Clergue owned a pulp and paper mill. In 1922 he described what he found and the renovations he made to the blockhouse in an article published in the *Sault Daily Star*, dated July 7, 1922. When Clergue acquired the blockhouse, the only parts that remained were the lower stone foundation and the walls. The site was surrounded by heaps of large boulders and stumps where the original stockade had been. The stumps could not be removed, so Clergue incorporated them and the boulders into a fence surrounding the property.[165]

The building originally consisted of one large room with gun loops measuring 6 inches square built into the wall. The lower level measured 22 x 28 feet and was constructed of unworked fieldstone and red mortar. The second and third levels were cantilevered over the first floor walls measuring 39 inches on each side; thus the second- and third-floor timber portion of the building measured 29 x 35 feet and 31 feet high. The roof was a pyramidal hip roof. A set of steps allowed access to the second floor and a suspended staircase to the third. The second floor was divided in half by a hallway with two large rooms on one side and three smaller rooms on the other. Clergue divided the two large rooms with cedar logs and probably created the division of the three smaller rooms as well. The third floor contained built-in storage chests. Clergue built a brick wall to form a small reception room in the front and a kitchen in the back. The original small iron door was replaced with a larger oak door, and windows were inserted into the stone walls, which were made from unhewn fieldstone. The gun loops were covered with permanent shutters. The second floor had to be entirely reconstructed. Clergue used cedar timber cut so that the logs would extend over the first floor walls. His intention was to recreate a blockhouse that would resemble those from the time of the French and Indian Wars. Based on the location of the chimney, the fireplace appears to have been located at the rear of the house, probably where the kitchen was eventually located.[166]

---

[164] Sault History Online, http://www.cityssm.on.ca/library/Clergue_Block.html; https://www.sootoday.com/columns/remember-this/remember-this-a-bachelor-retreat-of-the-proper-architectural-style-181639; http://www.historicplaces.ca/en/rep-reg/place-lieu.aspx?id=5595.
[165] Sault History Online, http://www.cityssm.on.ca/library/Clergue_Block.html; https://www.sootoday.com/columns/remember-this/remember-this-a-bachelor-retreat-of-the-proper-architectural-style-181639; http://www.historicplaces.ca/en/rep-reg/place-lieu.aspx?id=5595.
[166] Sault History Online, http://www.cityssm.on.ca/library/Clergue_Block.html; https://www.sootoday.com/columns/remember-this/remember-this-a-bachelor-retreat-of-

Eventually Clergue constructed a proper house for himself, his brother, their parents, and their two sisters in 1902. Clergue continued to use the blockhouse as an office until he left Sault Ste. Marie. The blockhouse then became the residence for the night watchmen of the paper mill until it was damaged by fire in 1974. In 1979 the blockhouse was declared to have architectural and historic value by the city council and was protected under Part IV of the Heritage Act. In 1995, St. Mary's Paper announced it needed that property for expansion. As a result, the blockhouse was moved to its current location in 1996.[167]

Figure 6.62. Clergue Blockhouse. *Left*: adapted from Fortwiki, http://fortwiki.com/Clergue_Blockhouse. *Right*: adapted from Wikipedia, https://commons.wikimedia.org/wiki/File:Clergue_Blockhouse_3.JPG.

Lacolle Mills Blockhouse

This blockhouse is located in Saint-Paul-de-I'Île-aux-Noix, Quebec, overlooking the Lacolle River, and was constructed ca. 1778–1781 as part of the British defense network to block American troops who used that river and route as a conduit to Quebec (Figure 6.63). It also protected a sawmill and lighthouse built on the river. The blockhouse was used again during the War of 1812 when ca. 4,000 American troops commanded by Gen. James Wilkinson attacked 600 British troops at Lacolle Mills on March 30, 1814. Soon after the war ended, the blockhouse was abandoned.

The blockhouse originally served as an outpost and strong point blocking access to Canada by American troops. It is square in plan and constructed of a stone foundation with a two-story wooden superstructure. The top floor is cantilevered over the first, and the roof is pyramidal. It also provided shelter for British troops transporting food and ammunition to

---

the-proper-architectural-style-181639; http://www.historicplaces.ca/en/rep-reg/place-lieu.aspx?id=5595.

[167] Sault History Online, http://www.cityssm.on.ca/library/Clergue_Block.html; https://www.sootoday.com/columns/remember-this/remember-this-a-bachelor-retreat-of-the-proper-architectural-style-181639; http://www.historicplaces.ca/en/rep-reg/place-lieu.aspx?id=5595.

garrisons in the area. The walls were made of large square-hewn timbers set one atop the other and dovetailed at the corners. Loopholes and embrasures allowed occupants to fire at the enemy from the safety of the building. The pavilion roof was supported by an impressive timber frame. There was a central fireplace, made of stone and brick, on the ground floor, sitting on top of an impressive sub-floor platform. This was one of twenty-five contemporaneous structures in Quebec. As with so many blockhouses, it became a residence, and was inhabited well into the early 1900s.[168]

Figure 6.63. Lacolle Mills Blockhouse, Quebec. Adapted from *Left*: Wikipedia ("Lacolle Mills Blockhouse1" by Blockhouse321 at the English language Wikipedia. Licensed under CC BY-SA 3.0 via Commons - https://commons.wikimedia.org/wiki/File:Lacolle_Mills_Blockhouse1.jpg#/media/File:Lacolle_Mills_Blockhouse1.jpg). *Middle*: Wikipedia ("Lacolle Mills Blockhouse3" by Blockhouse321 at the English language Wikipedia. Licensed under CC BY-SA 3.0 via Commons - https://commons.wikimedia.org/wiki/File:Lacolle_Mills_Blockhouse3.jpg#/media/File:Lacolle_Mills_Blockhouse3.jpg). *Right*: from the Saint-Paul-de-l'Île-Aux-Noix tourism website (http://www.ileauxnoix.com/eng/tourisme/blockhaus.html).

Based on this brief survey, key architectural features include thick, solid foundations and first-floor walls, a layout that is square with a central chimneystack, and proximity to a passageway that needed to be defended or blocked. Structural features such as a doorway, windows, a cantilevered second floor, and a pyramidal roof can be compared with Elkins' description of the Bayley-Hazen Road blockhouses.

However, the archaeological data from the Greensboro site provides no evidence of these features. Nevertheless, several features that were discovered at the Greensboro blockhouse site appear to correspond with those mentioned above. There are four major characteristics. First, the wall and foundation found in Probes 1 and 2, respectively, are substantial and consistent with those of other blockhouses. Second, there is an exterior earthen berm, or rampart, which may have been added for additional protection and/or strength. Third, the layout of the structure, including the distance between the wall/joist found in Probe 1 in 2011 and

[168] The Genealogy of Benjamin Bullock, The Lacolle Blockhouse, http://www.bullockgenealogy.org/features/p.2LacolleBlockhouse.php.

the foundation wall of Probe 2 is consistent with the overall measurement of contemporary blockhouses, as is the central location of what is considered the chimneystack in Sq. NE4. Finally, the site's proximity to the Bayley-Hazen Military Road would confirm the structure's function as militaristic and defensive. This comparative analysis supports the Greensboro blockhouse site and its interpretation as a blockhouse.

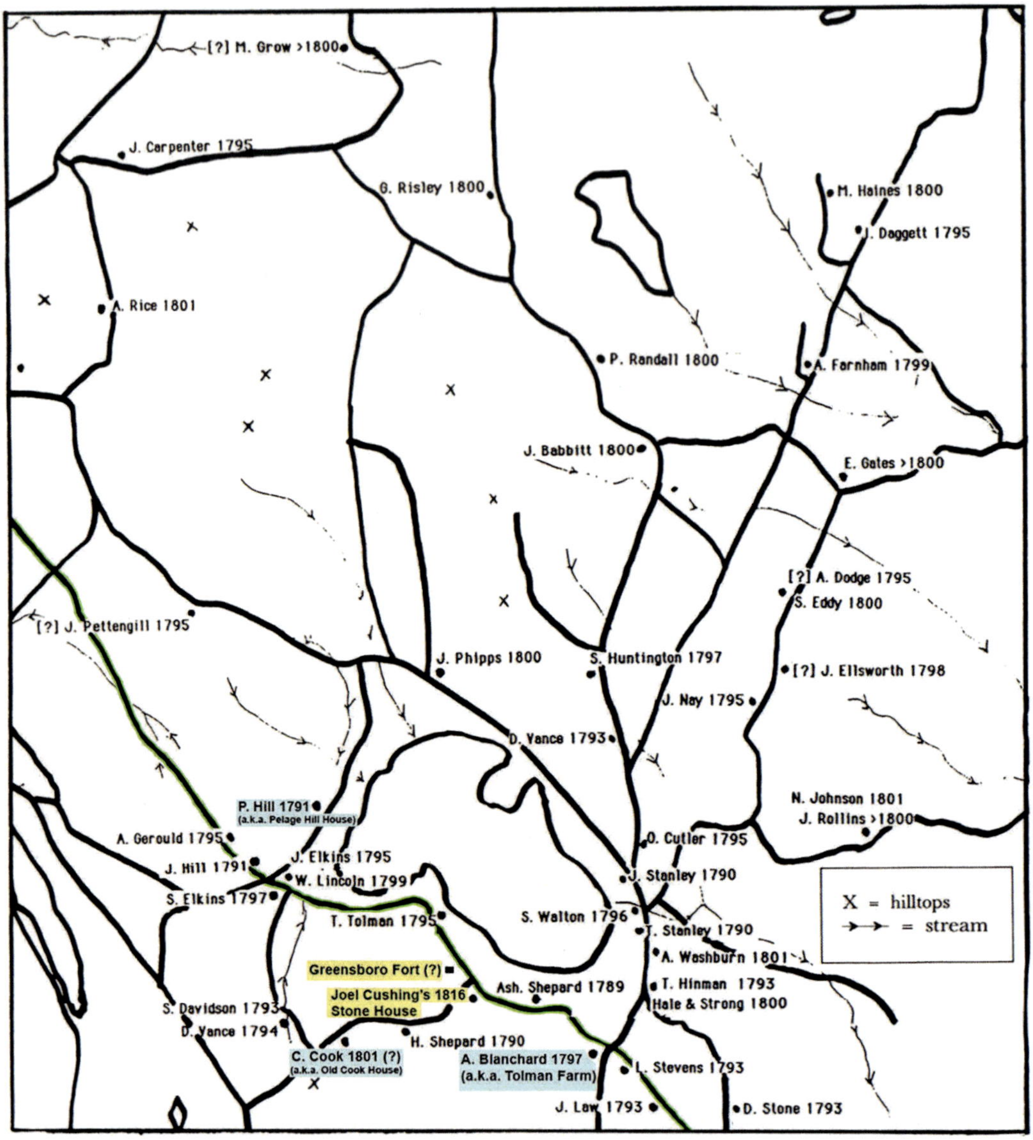

Figure 7.64. Interpretive map of Greensboro's early settlers (1789-1810) based on Rev. James P. Stone's *History of Greensboro*, 1854. Originally compiled by P. D. Watson and W. Smith for the *History of Greensboro, VT: The First Two Hundred Years*, (1990:24). Bayley-Hazen Road highlighted in green. The Hill, Blanchard, and Cook homesteads highlighted in blue can be cross-referenced with Figure 7.66, which includes pictures of the early houses. The Joel Cushing Stone house and Greensboro Fort are highlighted in yellow. Map adapted for use here to show the Joel Cushing Stone House and the proposed site of the Greensboro blockhouse. Used with permission of the Greensboro Historical Society

Chapter 7

Greensboro's Earliest Neighborhoods

Envision Greensboro as a wooded wilderness in the spring of 1789 when two Shepard brothers, Aaron and his younger brother Ashbel, came to the area with their families, having departed Newbury (then called Coos), Vermont, and traveled up the rough trail that was cleared in 1776 by Gen. Jacob Bayley before it was extended in 1779 by Col. Moses Hazen and his men. This was frontier territory. The Shepard brothers were thirty-one and twenty-six years, respectively. Aaron had one very young child (Jason), and both of the brothers' wives were pregnant during the journey. It is said they moved into the abandoned blockhouse on the western shore of Caspian Lake, an already-established shelter in the area, having transported their few belongings by hand sleds. In August, Aaron left Greensboro and returned to Newbury for the winter, but Ashbel and his family spent the winter in Greensboro, with their closest neighbor six miles away. The Shepards' foothold in the blockhouse and near Caspian Lake marked the beginning of settlement in this area.

This chapter will attempt to reconstruct Greensboro's earliest neighborhood from 1789 to about 1815 in the southern portion of town along the Bayley-Hazen Road, which was used as a migration route after peace was established between the United States and Great Britain in September 1783 (Figure 7.64). Due to the fires of 1832, which destroyed the earliest town records, supporting evidence will be taken from reconstructed land records, maps, transcribed church records, and extant treasurers' records, as well as resources found in the Vermont State Archives and Records Administration, Vermont Historical Society, and the Greensboro Historical Society.

Few of the reconstructed land records at the Greensboro town clerk's office described dwellings, boundaries, or farms, but all described land with acreage, lot and range numbers, the name of the original proprietor, and sometimes a specific reference point such as "the Burying Ground across from Capt. Crain's house" (now known as Lincoln-Noyes Cemetery), "Greensboro Pond," or "Hazen's Road." These were then plotted on the 1788 Lotting Plan with proprietors' names, but it could not be determined from the limited data exactly where the dwellings were on each property. This was especially true when a person held several properties. In the absence of the primary books of early land records, the Reverend James Stone's 1854 *History of Greensboro* provided approximate dates of settlement for the early families.

The interpretive map included in the 1990 *Greensboro History*[169] (Figure 7.64, adapted for use here) indicates where other early families settled around the lake and town, and it also provides their approximate arrival dates. The earliest arrivals appear to have been clustered between the present Tolman's Corner (Figure 7.66, below) and what is now known as Campbell's Corners at the north end of the lake.

Peleg Hill, a proprietor and early settler in 1792, established his home just around the corner on the North Shore Road. From the interpretive map, the southern quadrant at Tolman's Corner could include the Col. Levi Stevens house at or near where the John Allen house once stood opposite the first Hazendale farm stand.

Additionally, when comparing the 1800 and 1810 censuses for the order in which the families were listed on this portion of the road, they are virtually identical, along with other, lesser-known families listed who apparently did not stay in town or who moved about frequently. Cellar holes or other ruins have never been known to exist between the other recorded family locations on this section of the road. Using Rev. Stone's 1854 history and the plotted locations on the interpretive map (Figure 7.64), the families included Shepard, Ring, Tolman, Lincoln, Stevens, Rand, Cook, Durkee, and Cushing. Other families listed in the 1800 and 1810 censuses in the area between Tolman's Corner and Campbell's Corner include Aaron Ferren, John Law, David and John Vance, Amos Blanchard, Salmon Elkins, Benjamin Sinclair, Haren Burbank, Salmon King, and two Eastman families.

During the earliest migrations to the area, the four forts located on the Bayley-Hazen Road at Peacham, Cabot Plain, Walden, and Greensboro were the only structures that provided temporary shelter for individuals and families who needed a place to stay before they reached their destination. At times, these structures must have been periodic hubs of

---

169 Weber p. 24.

activity, with new families moving in and out. Some families journeying through the area built their own log homes, which eventually became way stations as well. The presence of new structures and temporary stay-overs for travelers provided much needed social occasions, especially for women, and were a focal point for conveying news.[170]

These early families set out plotting plans, cleared the land for farming, established neighborhoods, and built houses that remain recognizable today. To create a sense of who these people were as well as their place of origin and genealogy, we will take a closer look at plat records to determine the location of their homesteads.

In our research, we also found probate records to be a valuable source of information. Some residents possessed unusual personal items, such as Charles Cook Sr.'s "2 buffaloes @$14 and 2 buffaloes @$2.00," Peleg Hill's gun and bayonet (possibly from the Revolutionary War), and being the only person to own an umbrella. Two people owned silver and/or gold watches, which were considered very valuable at that time and a sign of prosperity. Some were owners of stock in the Greensboro library, one of the state's early library societies as enacted by the Vermont Legislature November 6, 1800. Such inventories reveal much about how they may have lived, and sometimes their occupations. Probate records also reveal whether the deceased was wealthy, poor, or insolvent at the end of life. Equally important is the value of their property and its location.

The Shepard Families

The Shepard family and their friends from Newbury exemplify how and when people relocated to new regions and established new towns. In part, their motivation was to acquire cheap land after the Revolutionary War. Land records from Newbury, Vermont, (1:43 and 1:108) confirm the local lore that in 1784 Aaron Shepard purchased acreage in Newbury and came from Litchfield County, Connecticut. Those properties were sold in 1790, after which he appears in the 1790 census (actually taken in 1791) at Greensboro. The third brother, Horace, and his young family accompanied Aaron and his family when they returned to Greensboro in the spring of 1790.

To set the stage for the Shepard families' exodus to Vermont, it is important to know that their parents were Aaron Sr. (1739–1781) and Susannah (Chamberlain) Shepard. Susannah was the daughter of Deacon Moses Chamberlain and Jemima Wright. Susannah was born September 9, 1740 in Orford, New Hampshire, and later went with her parents to Litchfield County, Connecticut, where the Shepards resided. One could surmise that this connection to the Newbury (Coos), Vermont, and Orford

[170] Stone 1877:209–215.

and Haverhill, New Hampshire, areas enticed Aaron Shepard Jr. to venture to this Connecticut River area in the 1780s. Two other reasons may have been his exposure to the Greensboro area during the Revolutionary War and his work surveying with James Whitelaw, the second surveyor-general of Vermont.

Research conducted by Gail Sangree in Litchfield, Connecticut, has revealed an early grantee deed wherein Aaron Shepard Sr. and Samuel Barnard together purchased land the same day that Aaron married Susannah Chamberlain, November 1, 1759. Aaron Sr. died in Woodbury, Connecticut, on April 10, 1781, but no burial place has been found. Susannah later married Col. Simeon Stevens and then James Corliss. Despite life's hardships and her frequent moves, Susanna Shepard lived to be over 100 years old. She died on October 4, 1840 in Greensboro, Vermont.

Aaron Shepard, Jr.

Aaron was born about 1760 at South Farms/Morris, Litchfield County, Connecticut. He married Phebe Andrus/Andrews about 1786 in Newington, Connecticut. After his father died in 1781, Aaron, Jr. became the family patriarch.

In 1777, military records list Aaron Jr. as serving in Litchfield County, Connecticut as a fifer in Capt. Amos Barns' Company in Col. Noadiah Hooker's Regiment. Two years later, in 1779, Aaron and Ashbel were both in Newbury, Vermont, serving in Capt. Simeon Stevens' Company of Col. Bedel's Regiment. Aaron also served as a scout under Capt. Frye Bayley.[171] This means that Aaron and Ashbel were already familiar with Newbury, Vermont, and the region prior to moving there with their families in 1784.

It is likely that since Aaron Shepard Jr. had been to the Greensboro area during his military excursions and as a surveyor, he knew the blockhouse/fort was unoccupied, which is why he chose this location for his new home prior to bringing his family there. Aaron Shepard Jr. met James Whitelaw of Ryegate, Vermont, and together they surveyed and established the Greensboro town lines in 1788 and subsequently the lotting lines for the town's proprietors. Aaron Shepard's field notes can be found in the Surveyor General's papers at the Vermont State Archives and Records Administration in Middlesex. The Aaron Shepard Jr. family was one of five families listed in the 1790 census as residents of the town, along with Paul Davison, Timothy Stanley, and Aaron's brothers Ashbel and Horace. Aaron Jr. was elected as the town representative to the Legislature in 1796, and was the enumerator of the 1800 Federal census for Greensboro. His barn served as the first school in Greensboro.

---

[171] Goodrich 1904:371.

Oddly, neither Aaron Jr. nor his wife Phebe were listed on the original list of members of the Congregational Church, dated November 24, 1804, nor were they included on the "List of Persons who Subscribed to the Covenant" in 1810; yet, their deaths were recorded in Congregational Church death records. Aaron, Jr. died May 1811. This is the only record of Aaron Shepard Jr.'s death date. Phebe (Andrus) Shepard, born August 10, 1766 at Newington, Connecticut, died after the 1840 census at the home of her daughter Phebe, who later married John Finch, at Meadville, Crawford County, Pennsylvania. No burial place for Aaron, Jr. has been found, though it is assumed he was buried in the Baker Cemetery, since Ashbel and Mary and other family members are buried there, or perhaps on the property of his farm.

After his death, Aaron Jr.'s property and possessions were administered by the Orleans County Probate Court in a document dated June 3, 1811: "It is the wish of the widow and friends of the deceased to signify that Thomas Durkee of Greensboro is administrator of the estate." The inventory appraisers are David Stone, Willard Lincoln, and Joseph Glidden. Their findings state that the real estate, "the farm," is valued at $600.00. The personal estate includes one yoke of oxen four years old, five cows, two calves, two heifers, a saddle and bridle (but no horse), eighteen sheep and lambs, and two swine as the farm animals. Plow irons and chains are listed along with an axe and earthen milk pans, but no other farming equipment. Dishes, two wine glasses, tumblers, two iron candlesticks, and other utensils make up the kitchen and household equipment. Furniture was comprised of a cherry wood table, a breakfast table, ten kitchen chairs, four bedsteads, and one wooden clock. Clothing consisted of one coat, one great coat, cotton stockings, one pair of pantaloons, and vests. There were several sets of bedding, towels, and tablecloths. Like his brother Ashbel, he had an old loom and rigging, a foot wheel, and a large spinning wheel, but "no supply of wool as did Ashbel." His case of protracting instruments and surveyor's compass and chain are listed. Total inventory of personal estate was $863.81 as entered with the court on June 1, 1811 by the appraisers.

The list of claims against the estate includes a note to his mother, Susannah Corliss, at $104.00 and a note to the Meridian Sun Lodge #20 (formerly #17), F. & A. M. at Craftsbury, whose roster included Samuel C. Crafts and Horace F. Graham as its most prominent members. Eighteen other claimants are listed with notes or on-account amounts totaling $419.56.

The administrator, Thomas Durkee, tried to sell the farm by private sale by advertising in the *North Star* newspaper in Danville, but was unsuccessful. The "widow's dower" or "thirds" also had to be provided out of the sale of the real estate. Hon. Royal Corbin, Probate Judge, decreed that Durkee must sell the farm at public venue, the sale of which was presented to the judge on the first Monday in September 1813. The buyer was Joel Cushing of Montpelier who paid $375.00 for the farm, far less than the appraised value of $600.00. The widow was paid $118.96 in furniture, provisions, and for support of the family.

A descriptive list of Aaron Jr. and Phebe Shepard's children can be found in the *History of Greensboro* genealogies.[172] The following is a brief summary of their children.

1. **Jason**, born July 31, 1788 in Connecticut. Jason made the trek from Newbury Vermont to Greensboro; married Alice Smith, daughter of Amos Smith Sr., a Revolutionary War soldier with service from Connecticut. Alice died in Greensboro on May 26, 1811 at age twenty-one, just fifteen days after her father's death. Jason later married Hannah Skinner of Albany, Vermont. Jason died September 9, 1834 at Newburg, Ohio.
2. **Prentice,** born August 30, 1789, married Esther (family name unknown). They are said to have settled in Washington County, New York.
3. **Aaron,** born March 15, 1791 in Greensboro, married Anna Bennett; died August 4, 1855 at Newburg, Ohio.
4. **Clarinda,** born February 7, 1793 in Greensboro.
5. **Achsah,** born August 11, 1796 in Greensboro; died March 4, 1813.
6. **Phebe,** born January 3, 1798 in Greensboro; married John Finch of Meadville, Pennsylvania.
7. **Caroline Matilda,** born January 1, 1809 in Greensboro.

Ashbel Shepard

With his older brother Aaron Jr., Ashbel went to Newbury from South Farms, Litchfield County, Connecticut in 1784. He had served in Capt. Simeon Stevens' Company in 1779; in Col. Peter Olcott's regiment out of Newbury in 1781; and as a scout in Capt. Frye Bayley's Company, all of which was confirmed by the payroll lists from 1779 to 1781.[173] He married Mary (Vance) Barrett on September 28, 1786 in Newbury, Vermont. According to land records, Ashbel also owned land in Newbury before he and Mary moved to Greensboro. Their first child, William Scott Shepard, was born March 25, 1790 in Greensboro, the first non-Native American born in town.[174] To celebrate this momentous event, the town proprietors voted to give him 100 acres of land in Lot 9 in the 14th Range. This family

[172] Weber 1990.
[173] Goodrich: 628.
[174] Stone 1877:212.

was the first to spend a lonely winter in Greensboro with the closest neighbors six miles away. It is likely that Ashbel and Mary first lived with Aaron's family in the old fort before purchasing their own property on the corner of Randolph Road and the Bayley-Hazen (now Lake Shore) Road, later known as the Durkee and (Ezekiel) Rand farm. Ashbel and Mary were in the original membership list of the Church of Christ in Greensboro in 1804, and Ashbel was the first town moderator. Chester Brown, whose first wife, Calista Shepard, was a descendant of the Shepard family, published a history of the family in 1894 when he resided in East Hardwick.[175] Brown tells how, in March 1790, the Cutler family and closest neighbors in Craftsbury "were reduced in provisions to one quart of Indian meal, the snow was four feet deep, and they were making up their minds that they must starve to death when Ashbel Shepard providentially killed a moose, dressed it, and the next morning he took a quarter of the meat on his shoulder and carried it on snowshoes the six miles to the Cutler family. Mrs. Cutler met him at the door and said, 'Well, I never will give up hope again!'"[176] This journey occurred in the same month that Mary Shepard gave birth to her first child William.

Ashbel died in Greensboro on June 4, 1808 at age forty-five, and Mary seven months later on January 9, 1809. Both are buried in Baker Cemetery (Figure 7.65).

Figure 7.65. Ashbel and Mary Shepard's gravestone in the Baker Cemetery in Greensboro, Vermont. Photograph by Jill L. Baker, 2016.

[175] Brown 1894:10.
[176] Brown 1894:10.

Estate papers for Ashbel were filed in Craftsbury at the Orleans County Court, with Ebenezer Crafts, Esq. as Judge of Probate. The intestate petition was presented by William Sanborn and Mary Shepard, Ashbel's widow, as joint administrators because Mary was "in feeble health weakened by the long and exhausting sickness of her husband."[177] The inventory includes two cows, a heifer, and a calf; a foal, old colt, and old red horse; seventeen sheep, a hog, old plow, sleigh, horse sled and farm implements; an old cart, four beds and cotton and woolen bedding; little clothing; three more old bedsteads with cords; six dining and six plain chairs, a breakfast table; a chest of drawers; kitchenware, warming pan, a gun, side saddle, a loom, great and little spinning wheels, thirty pounds of wool, baskets, and utensils for weaving. The appraisers were Col. Levi Stevens, Ashbel Hale, and Willard Lincoln. The real estate was appraised and described as being in the southeasterly half of Lots #6 in the second and third ranges, supposed to contain one hundred acres amounting in the whole to $800.00, one third of which is $266.67. As the widow's portion, "this third was then set off to Mary Shepard the widow for her dower or thirds, which consists of the southwest room in the house . . . together with the chamber directly over it and an equal privilege in the kitchen, cellar, and back rooms; also the bay and stable in the south end of the barn, with an equal privilege on the floor and yard adjoining," dated December 16, 1808.[178]

About three weeks later, Mary died on January 9, 1809, and the estate was declared insolvent by William Sanborn, who suggested that commissioners be appointed to receive claims against the estate. The widow had received "her thirds," and an amount was granted for the support of their children. Guardians were appointed for the children. All of Ashbel and Mary's children were born in Greensboro:

1. **William Scott,** born March 25, 1790. In 1854 he was said to be "of Southport, Wisconson."
2. **Lucinda.** No information available.
3. **Ansel,** born March 4, 1794. On September 28, 1828, he married Elizabeth Cole Richmond in Woodstock, Vermont. He died there April 21, 1858.
4. **Harvey,** born 1799, married Amanda Adams on January 29, 1824, in Hardwick.
5. **Mary.** No information available.
6. **Ashbel,** baptized February 4, 1805, died February 14, 1813, in Greensboro.

---

[177] Orleans County Probate Court 1:72.
[178] Orleans County Probate Court 1:72.

Horace Shepard

Horace Shepard, the third Shepard brother, was born in 1765 (month and day unknown) in South Farms/Morris, Connecticut. He went to Newbury, Vermont, in 1784 with his family. He married Martha Ring of Haverhill, New Hampshire, on May 10, 1787, in Haverhill, and they had eight children. Martha was the daughter of Jonathan Ring (1736–1815) and his first wife. Horace died on March 2, 1829, in East Hardwick, and Martha died less than a year later in January 1830. Both are buried in Center Cemetery, Hardwick.

Horace did not move to Greensboro with Aaron and Ashbel in the spring of 1789, but relocated there from Newbury one year later in March of 1790 when Aaron and his family returned to Greensboro. The first six children were baptized December 28, 1805, according to the Congregational Church Records of Greensboro. These children were Zilpha, Milo, Jonathan, Parmela, Sally, and Hannah. In August 1806 they were "dismissed from the Congregational Church to the Baptist Church" in Hardwick, and appear in the 1810 and 1820 censuses in Hardwick. Rev. Stone's history claims that this family lived on what later became known as the Ring farm, which was on the road from Merrill Williams (living in the Stone House) and Mr. (Charles) Cook, later known as the Cook Hill Road.

Susan Shepard

The fourth child of Aaron Sr. (1736–1781) and Susannah (Chamberlain) Shepard was Susan born 1768 in South Farms/ Morris, Connecticut. In 1784 she traveled with her siblings to Newbury, Vermont, after the death of their father in 1781. She married Levi Stevens November 29, 1790, at Newbury. Levi was the son of Col. Simeon Stevens, a first settler of Newbury, grantee of both Newbury and Haverhill, New Hampshire, and a Revolutionary War soldier with service from New Hampshire. Simeon married second, Susan Shepard's mother, Susannah.

Levi Stevens was the succeeding owner of Timothy Stanley's gristmill in the village. Colonel Stevens began on the farm owned in 1854 by C. B. Field according to the Stones' history. Susan and Levi Stevens had three daughters, Matilda, Sally, and Mary, all of whom died at a young age. Two died in the fall of 1802 along with their mother Susan of a dysentery epidemic in town. They are buried in Baker Cemetery. In 1854, Rev. Stone wrote, "Suffering and sorrow — In the year 1802 when of the seven families, 14 persons were suddenly removed from earth by Dysentery," followed by smallpox. "These were the wife and three children of Col. Levi Stevens (wife nee Shepard), three children of William Sanborn, two of Timothy Stanley, two of Joseph Stanley, one of Capt. David Stone, one of Capt. James Andrews, and one of Stephen Adams. Mrs. Stevens was the second adult who died among the settlers."

Azubah Shepard Sr.

Azubah was the last of the five children of Aaron Sr. and Susannah Shepard. Born in 1772 in Morris, Connecticut, she journeyed with her siblings to Newbury about 1784. She married Obed Cutler, son of Nathan Cutler. Azubah passed away August 31, 1806, "leaving an infant daughter two days old," according to the Congregational Church records. They appear in the 1800 census of Greensboro with one male under age 10, one male age 16–26, three females under age five, and one female age 16–45 (Azubah), and in the 1810 census Obed appears with his second wife and their children.

Tolman Farm (Four Corners)
(a.k.a. A. Blanchard 1797)

First Cook Cabin, Cook's Hill
(a.k.a. C. Cook 1801)

Old Cook House, Cook's Hill
(a.k.a. C. Cook 1801)

Pelage Hill House
(a.k.a. P. Hill 1791)

Caspian Lake from Cook's Hill Field
before 1941

Caspian Lake from Cook's Hill
after 1941

Figure 7.66. Photographs cross-referenced with interpretive map in Figure 7.64 above. Top two and middle left photographs taken of originals in 1977, housed at the Greensboro Historical Society, used here with permission. Middle right photograph of the P. Hill house taken by Pat Haslam in 2013. Lower two photographs taken in the 1940s/1950s of Caspian Lake from Cook's Hill. Lower right photo likely taken after 1941 because the monument to Bliss and Sleeper appears. Lower two photographs courtesy of J. Bascom.

Second Wave of Settlers

In this section, we will discuss the second wave of settlers who built their cabins along the western shore of Caspian Lake on the Bayley-Hazen Road prior to 1810, including Thomas Tolman Jr., Thomas Durkee, Jonathan and David Ring, Willard Lincoln, Charles Cook Sr. and Joel Cushing (Figures 7.64 and 7.66).

This neighborhood journey will begin on the Hardwick/Greensboro town line, just southeast of Tolman's Corner on the Bayley-Hazen Road, and proceed northwest to the settlement at what is now known as Campbell's Corner. For relevant dates, refer to Rev. James P. Stone's valuable *History of Greensboro* (1854) wherein he listed the earliest settlers and their approximate dates and locations in town, as provided to him by Grandma'am Stanley, wife of Joseph. This history can also be found in the *Vermont Historical Gazetteer* by A. L. Hemenway, Orleans County portion, Volume III (1877). The list of pioneers and their years of settlement is projected on the interpretive map on page 24 of *The History of Greensboro: The First Two Hundred Years,* published by the Greensboro Historical Society in 1990. Corrections to this map have been made for locations of Aaron Shepard and Joel Cushing (Figure 7.64).

Our journey begins at the Hardwick/Greensboro town line on the Bayley-Hazen Road/Hardwick Street.

#1 John Law,[179] born in Putney, Vermont, and died in 1815 in Fletcher, Vermont, was said to have settled in Greensboro in 1791 according to Rev. Stone. John initially located on what later became the property of Charles Cook,[180] soon after which he relocated to almost the Greensboro-Hardwick town line on what is Hardwick Street at the time of this writing. However, he was not listed in the 1790 (1791) federal census of Greensboro. He and his wife attended the wedding of Joseph Stanley and Mary Gerould on July 11, 1793. No information can be found about John Law in reconstructed land records, cemetery, census, or Vermont State vital records. However, he was listed in Capt. Elijah Gates' Company of Volunteers in Col. Benjamin Wait's Battalion of the State of Vermont from July 17 to December 15, 1781. There are other familiar names in this roster, especially that of Aaron Shepard, so it seems very likely that this is the same John Law who came to Greensboro at this time.

---

179 It should be noted that the numbers correspond to the directional route shown on the interpretive map, Figure 7.64.

180 Stone 1854:12–13.

#2 Levi Stevens settled in town before July 1793, not far from John Law, on what is now Hardwick Street. Levi was a colonel in the Vermont militia and later succeeded Timothy Stanley as the owner of the gristmill in the village. His first wife was Susannah Shippen (*sic, Shepard,* sister of Aaron, Ashbel, and Horace Shepard), whom he married November 29, 1790 at Newbury. She died of dysentery on September 26, 1802, along with two of her daughters, when an epidemic spread through town. Levi Stevens had two other marriages: Esther (family name unknown) who died before 1810, and Sally (Johnson) Brown, whom he married December 17, 1845, at Greensboro. This may have been the same Levi who also served in the War of 1812 in Capt. Samuel Gordon's Company, 11th Regiment. In the 1815 Orleans County General List, Levi owned 138 acres of land, two houses, and one outbuilding, a total value of $2,500.

Simeon Stevens and his first wife Sarah Hadley were the parents of Levi.[181] After the death of his first wife, Simeon later married Susannah (Chamberlain) Shepard, mother of Aaron, Ashbel Shepard and siblings, making her the stepmother of Levi Stevens.

#3 Amos Blanchard is said by Rev. Stone to have come to Greensboro about 1797, locating first at what we now know as Tolman's Corner, and later relocating on the road to Baker Cemetery and Barr Hill. He was a New Hampshire native born in Canterbury in 1773. He married Betsey Tolman, date and place unknown; the names of her parents are also unknown. He died in Greensboro March 10, 1829, age fifty-six, after Betsey had died just two months earlier on January 10. Both are buried in Baker Cemetery. They had no children, yet the 1800 census lists Amos' household as having four other additional members: a male under age 10, a male aged 10–15, a female under age 10, and a female 16–25. They could have been children who did not survive, relatives, or farm labor and domestic help.

Proof that he owned the property at Tolman's Corner (Figure 7.64, 7.66) can be found in Greensboro Land Records, Book B:225, in which the heirs of Amos Blanchard (his sisters, Martha Cutler, wife of Nathan Cutler Jr.; Keziah, wife of Obadiah Glines; and Sally Emery of Chichester, New Hampshire) and two others, not named, are represented by Stephen Sherman of Greensboro and John Boardman of Glover in a quit claim deeded to Enoch Tolman, "meaning to convey all the right and title we have in the farm now owned by Capt. Aaron Walker and deeded to Walker by Enoch Tolman 15 Sept 1836." Tracking back from that transaction to the entry in Book A:520, which recorded that Enoch Tolman sold the property for $1000 to Aaron Walker of Peacham, Vermont, we find the following mention in the property description: "a

---

181 Wells 1902:697.

tract of land and farm situate in Lot 7 in 3rd range, and Lot 7 in 2nd range of which John Knight was the original proprietor." There are some exceptions about various parcels to Timothy Hinman, Ebenezer Strong, Ashbel Shepard, and Levi Stevens, "and except also the privilege to the South School District," dated in 1823 and re-recorded in 1835. This fits the present Tolman's Corner area on the overlay map of lots and ranges over the modern topographical map of Greensboro. The farm was later the showplace homestead of Henry S. Tolman, son of Enoch Tolman.

Amos Blanchard was one of the more prosperous of the early settlers in the neighborhood. His many possessions at the time of his death are listed in the inventory as well as financial demands against 117 men, relatives as well as townspeople, with thirty-one of these loans described as "doubtful." He was a bank! Perhaps it was because of the many uncollected loans that he died insolvent, as his total estate was "insufficient to pay all the just debts."[182] Amos owned two houses and one outbuilding valued together at $1400.00 on the Orleans County General List of 1815. His second house and farm on Barr Hill Road was apparently the one where he died.

At a Probate Court hearing held at Barton, Orleans District, March 18, 1830, Hon. John Kimball, Judge, "was presented to a writing"[183] to said Court signed by most of the named relatives of Amos Blanchard, Esq., late of Greensborough, requesting that John M. Scott of Greensboro be granted Administrator, and that Stephen Sherman and Salmon Elkins of Greensboro, and John Boardman of Glover, be appointed appraisers to take inventory. This inventory consisted of real estate described as "the homestead of the deceased being the farm on which he lived and died supposed to be about 110 acres estimated at $750."[184] There were almost four pages, legal-sized and double columned, of personal estate items that included two oxen, three cows, one horse, three live hogs, and eighteen sheep. Items not found in the inventories of other contemporaries include 250 pounds and five tubs of sugar; a silver watch valued at $10.00 and a gold watch at $15.00; a candle mold; a sausage machine; three tea chests, Japanned; several tea sets; a cradle bedstead; clock reel, foot reel, and a woolen wheel; a flax "sive," two pounds combed flax, and twelve pounds sheep wool. Blanchard held a small library of his own containing thirteen volumes, including gazetteers, a Bible and a Testament, *Williams Sermon*, *Trails of Virtue*, and dictionaries. He also held "two shares in the Greensboro Library at $4."[185]

Reference to the gazetteers and books was thought to be the first reference to a precursor of the Greensboro Library until the inventory of

---

[182] Orleans County Probate Court 2:102, 134–135, 291.
[183] Orleans County Probate Court 2:102, 134–135, 291.
[184] Orleans County Probate Court 2:102, 134–135, 291.
[185] Orleans County Probate Court 2: 291.

John W. Ellsworth, who died intestate in 1816 in Greensboro, listed two shares in the Greensboro Library at $5.00.[186] Three cows, one each, were kept at F. Detrayer's, R. Randall's, and M. Haines' farms. A second inventory added in July 1829 lists land holdings in Greensboro consisting of half of Lot #4 in 12th range, 50 acres; half of Lot #4 in 19th range, 70 acres; 120 acres in Lot #8 in 19th range, Marsh place; and 100 acres of Lot #5 in 15th range called Randall farm valued at $300.00. There is also "one crane in the South house," which could be the house at Tolman's Corner. In a third "additional inventory," there are seven numbered lots of land in the Town of Woodbury.[187]

#4 Ashbel Shepard went to Greensboro with his brother Aaron in spring 1789. He and his wife spent the first winter on the Bayley-Hazen Road . They were the only family in town. Brother Aaron and his wife Phebe had returned to Newbury for the winter.

#5 The Stone House location (Figure 7.67). The interpretive map on p. 24 of the *History of Greensboro* [188] describes (in error) this location as the site of Aaron Shepard's settlement in 1789. In Figure 7.64 above, this has been corrected. According to Stone's history (1854) and now legend, Aaron is said to have moved into the old blockhouse in spring of that year. Some have incorrectly considered the Stone House to be the blockhouse, but it is not. It has been established that the Stone House was built in 1816, and the old blockhouse was built in 1779.[189] The blockhouse was thought to have been built on the Bayley-Hazen Road just north of the Cook Hill turn, which we verified in the course of this archaeological investigation.

Figure 7.67. The Stone House on the Bayley-Hazen Road (present-day Lake Shore Road). *Left*: Taken possibly during the ownership of Henry Miller, 1903–1920. Courtesy J. Bascom. *Middle*: Taken in the 1920s or early 1930s. Loaned by R. Mercier. *Right*: Taken in July 2009 by Mary L. Baker.

[186] Ellsworth is not within the scope of this neighborhood study.
[187] Orleans County Probate Court, Vol. 2:101, 134–5, 291.
[188] Weber 1990.
[189] See Haslam 1996.

#6 Joel Cushing's house was located about half a mile north of Randolph Road. On the left side of the Bayley-Hazen Road sits the Stone House just before the rise of the hill where the granite Bliss-Sleeper monument is located. The Stone House could have been started by Joel Cushing before 1815, as evidenced by the Orleans County Grand List of 1815, which recorded that he owned a house valued at $1,000.00, a substantial house for the times as compared to the Shepard estate house valued at $475.00 in 1813. It was probably finished shortly after 1816. It is also possible that Aaron Shepard (who died May 1811) may have started to build the Stone House, and Cushing finished it. Whatever the scenario, Joel Cushing acquired land and buildings from the intestate estate of Aaron Shepard's widow Phebe Shepard and Thomas Durkee, administrators, on October 21, 1813.[190] This deed describes "110 acres of land and premises with all buildings" valued at $475.00, and would have included the portion of land now being studied as the probable site of the old blockhouse on the Bayley-Hazen Road near the curve of the Cook Hill Road.

A title search supports the extent of the property. The acquisition of the Shepard property by Joel Cushing on this date signals the possible start of the construction of the Stone House. It is not known exactly how long it would have taken to build a house of granite from one boulder on the property using hand tools and very basic methods of splitting and moving the huge blocks of granite, though it was likely over the course of several winters. In 1836 Joel Cushing sold his various Greensboro properties and moved to Hardwick, where he died in 1849 at age sixty-seven. He is buried in the Lincoln-Noyes Cemetery in Greensboro next to his wife Betsey (Talbott), who died after 1860 in Hardwick, and their children Horatio and Lorenzo, who both died before their third birthday.[191] At a tax vendue in Greensboro in 1835, Cushing acquired numerous properties.[192] At the time of his death, Cushing"s real estate holdings consisted of a farm in Hardwick of 160 acres with no value entered.[193] His wife Betsey survived him.

#6 Thomas[6] Tolman was said by Rev. Stone to have been in Greensboro by 1795.[194] His location was near the entrance of the Aspenhurst turn where the C. W. (Bill) and Pucky Carter cottage is now located. Thomas[6] Tolman (1756–1842), son of Thomas[5] born 1727–1821, has been well studied in past papers published in the *Hazen Road Dispatch,*

---

[190] Vol. A:451 Greensboro Land Records.
[191] Also a Luther Cushing, who is thus far unplaced.
[192] Book C:28.
[193] Caledonia Probate Court Vol. 19: 324, 326, 327.
[194] Stone 1854:13.

so he will be recapped more briefly in this sketch.[195] Thomas[6] was born in Needham or Attleboro, Massachusetts, the son of Thomas and Elizabeth (Pike) Tolman. He married in 1780 Lois Clark at Attleboro. They had eleven children. This couple is one of the few families, as original proprietors, to have settled in town. He died in Greensboro September 7, 1842, age eighty-six years. Both are buried in the Village Cemetery. Both were original proprietors on the Greensboro lotting plan of the original grantees, Lois Tolman being one of two women to be proprietors. The second woman is Susan Allen, presumed wife of Joshua Allen, also an original proprietor. The Allens never took possession of their property.

Thomas[6] served in the Continental troops of Massachusetts and Vermont in the Revolutionary War in Lt. Seth Warner's regiment and other units. His widow Lois survived him and received a pension on #W18146 and Bounty Land Warrant #2121-200 issued February 14, 1793. He held the rank of lieutenant and paymaster in that regiment. Congregational Church records show that he had been a minister earlier in his life,[196] and as early as 1787 the Reverends Messrs. Tolman and Wood camped on the shores of Caspian Lake and prayed fervently that future inhabitants might be virtuous.[197] Tolman was the chairman of the Selectmen in 1794. Later he served as one of the first town clerks from 1803–1812, and was a tax collector.

The Tolmans held six lots between them: Lois is named on Lot 1 in Range 6, Lot 4 in Range 12, and Lot 5 in Range 12. Thomas held Lot 5 in Range 16, Lot 6 in Range 16, and Lot 10 in Range 15. Yet none of their original six lots match Lot 5 Range 5 where they actually lived at the corner of the Bayley-Hazen Road and the turn to Aspenhurst cottages.

The 1815 *Orleans County General List* shows Thomas Tolman with 500 acres of land and no houses or outhouses. Three years later, in his application for a pension in 1818, he is in greatly reduced circumstances and declining health. In the declaration of his *Schedule of Whole Estate and Income*, he is described as living in "one small one-story house, 22 x 24 feet and 1/2 acre of land adjoining, one cow and one heifer . . . income I have none, except a contract I have with a son for firewood, at the door, unchopped, for a stove . . . and for keeping my two cows, and wool, potatoes, and apples for my family, but this contract is on condition to be fulfilled on my part, which I can, in no part fulfill without receiving my pension. My wife is age sixty-two years [and] of slender health." He

---

195 See his genealogy in the appendix to Weber, *The History of Greensboro: The First Two Hundred Years*; Haslam, "Aspenhurst Farm and the Tolman Connection: the 200th Anniversary of Thomas Tolman, Jr. to Greensboro" 1995, (*Hazen Road Dispatch* Vol. 20:35); and Sangree, "Two Letters of Thomas Tolman the Younger," 2003 (*Hazen Road Dispatch* Vol. 28:25).

196 Stone 1854:35, note.

197 Stone 1877:210.

describes their three unmarried daughters: Roxelana, age thirty-six, "an invalid"; Achsah, age thirty, "her constitution utterly destroyed by undue labour in the family"; and Maria, age twenty, "slender constitutionally for any hard labour." Yet, Roxelana and Achsah later married. There is no probate estate for Thomas Tolman.

#7 **Willard Lincoln** arrived in Greensboro about 1799. His dwelling site was near the present Lincoln-Noyes Cemetery, that portion of the Campbell's Corner hamlet. Lincoln succeeded Josiah Elkins on his farm. He migrated to Greensboro, Vermont, from Brookfield, Massachusetts, and brought his widowed mother Dorothy Lincoln with him in 1799. She passed away March 27, 1816, at the age of seventy-four. Willard Lincoln was one of the Greensboro residents who owned a house valued at $1,000.00 or more who was included in the 1815 General List for Orleans County. He then owned one house valued at $1,000.00 and no outbuildings. In Land Records Book A:29 there is a transaction that was re-recorded October 1, 1832 from Amos Smith Jr. to Willard Lincoln, for a consideration of $400.00 for a piece of land, the eastern half of lot #2 in the 13th range drawn to the original right of John May, excepting one acre, which he purchased from Jeremiah Lincoln, which was located southerly of said lot. This Jeremiah Lincoln is unplaced.

Willard was born August 8, 1779, in Brookfield, Massachusetts (*Brookfield, MA Vital Records to 1850*), to Loring and Dorothy (Mower/Moore) Lincoln. In 1817 he married Elizabeth Cook, the daughter of Charles Cook Sr. and Elizabeth Burbeck or Burdick. Willard and Elizabeth had nine children.

In an interesting side story, in 1812 Amos Porter II (born in 1763, and died in Greensboro April 29, 1815, at 52 years of age) of Farmington, Connecticut, came to Greensboro and settled in the western part of town. Amos Porter II had traveled extensively to China selling Vermont ginseng as told in his journal, *The China Journal of Amos Porter 1802–1803,* published by the Greensboro Historical Society in 1984. Willard Lincoln's sister, Dorothy Lincoln, married Porter in 1803 upon his return from China.

Willard Lincoln was the town tax collector in 1831 and a Representative to the Legislature in 1836. He held other town offices, and apparently was a respected member of the community. Mr. Willard Lincoln died in Greensboro August 12, 1839, at age sixty, "of disease of the heart."[198] Both he and his wife, Elizabeth (Cook), are buried in the Lincoln-Noyes cemetery; Elizabeth having died August 4, 1875, at eighty-one years of age. At a Probate Court at Irasburgh, the intestate estate of Willard

[198] Vermont Vital Records

Lincoln was presented August 28, 1839, and his widow Elizabeth was appointed administrator. The estate was represented insolvent. Three commissioners were appointed to substantiate claims. A summary of the inventory lists the real estate as the "Home Farm owned by Charles Cook,"[199] 100 acres in timber and pasture Lot #3 in 8th Range, and fifty acres of a timber lot in Albany (town), being Lot #96. The personal estate included horses, two oxen, cows and calves, seventy-four sheep (more than his neighbors), hogs and pigs; various items of household furniture, linens, one clock, "sundary books," two wheels and one reel, and loom. Kitchen furniture and equipment items were listed. Lincoln had harvested and stored 35 tons of hay, 150 pounds of cheese, 300 pounds of butter, 100 bushels of oats, and 400 pounds of potatoes. Farm equipment, two sleds, and a sleigh were also enumerated. He had a land claim in Sheffield (town) at $30.00. The inventory was recorded September 28, 1839 by Stephen Sherman and Charles Cook, appraisers.[200]

Salmon Elkins, according to Rev. Stone, came to Greensboro in 1797[201] and sited his cabin across from the Lincoln-Noyes Cemetery. At the time of this writing, this location is known as the Sylvia Lotspeitch property. Interestingly, neither Salmon Elkins nor any Elkins family member was located in the Greensboro Cemetery Census of 1980. Research in some Elkins family trees show that J. Elkins, #11 of this paper, was *Josiah* Elkins, brother of Salmon, sometimes spelled Solomon. Josiah and Salmon were the sons of Deacon Jonathan Elkins (1734–1808) of Peacham and his wife Elizabeth Rowell. Interestingly, another brother, Jonathan Elkins (1761–1852), wrote a memoir later in his life, already mentioned above, which was published by the Vermont Historical Society in their *Quarterly Proceedings*, no volume number, but dated 1921, p. 185–211, invaluable because it is one of the few surviving journals of that time.

Salmon Elkins was born April 11, 1771 in Haverhill, New Hampshire, and must have been acquainted with the Ring family, who were also from Haverhill. He is listed in Greensboro only in the 1800 census and is enumerated in the 1815 Orleans County General List with 108 acres of land and one house, total value $600.00.

In 1854 this property belonged to Matthew Marshall, and Salmon Elkins was not listed in any other censuses, yet he was living in Greensboro and served as an appraiser for many inventories of estates as late as March 1830. In a conveyance from Salmon Elkins of Potton, Province of Lower Canada, to Burrell Vance regarding a parcel of land in the northeasterly corner of Lot #3 in the 3rd range, it can be determined that Elkins left Greensboro sometime shortly after taking the inventory for Amos

199 Orleans County Probate Court 5:234.
200 Orleans County Probate Court Vol. 4:449.
201 Stone 1854:13.

Blanchard (1829) and, at the latest, the date of this deed August 22, 1831.[202] He died in 1838 in Potton, Canada East. After 1840, Potton became part of the lower Province of Quebec, Canada, just over the US border from North Troy, Vermont. A son, Ralph B. Elkins, listed in the Greensboro 1820 census, was still there in May 1834 when he granted property to Elnathan Strong.[203]

#9 James Hill arrived in Greensboro sometime around 1791 at about thirty years of age, and chose a location on the Bayley-Hazen Road, just north of the North Shore Road, for his homestead.[204] He married Mary Talcott before 1789. James is said to have moved to Canterbury, New Hampshire, sometime after the 1820 census of Greensboro according to the Lewis Hill papers, and possibly died there. The 1815 Orleans County General List includes James Hill with 199 acres of land at that time with one house valued at $460.00.

James Hill was born September 12, 1761 in Saybrook, Connecticut, son of Peleg[4] and Dorcas Hill. He was probably married in Connecticut since the 1790 Connecticut census lists James in Saybrook with a wife, a son, and two daughters. He appears in the 1800, 1810, and 1820 censuses of Greensboro.

#10 Asabel Gerould/Gerald arrived in Greensboro in 1795, according to Rev. Stone, locating on the Bayley-Hazen Road just north of Campbell's Corner. Asabel, sometimes spelled Asahel, is the brother of Mary Gerould, who married Joseph Stanley July 25, 1793, the first marriage in the town of Greensboro. Mary and Asabel were the only children of Joshua Gerould and Prudence (Scott) Gerould. Joshua died three years after their marriage in Sturbridge, Massachusetts, in 1776. Prudence became the third wife of Nathan Cutler (1736–1814) of Minden/Craftsbury. Asabel Gerould does not appear in any census, nor is he found in the Greensboro land or cemetery records. No name for his wife can be found, and no death date in Vermont Vital Records.

#11 Josiah Elkins is said to have come to Greensboro in 1795 and located on the Bayley-Hazen Road and the North Shore Road four corners. Mr. Willard Lincoln succeeded him, and in 1854 this farm was held by William W. Lincoln. He may have been in Greensboro before 1793 when he was one of several men who conveyed Mary Gerould, who

---

[202] Land Records Vol. A-18.

[203] Land Records. A:381.

[204] Stone 1854:11.

relocated from Sturbridge, Massachusetts, to Greensboro, Vermont in several phases.[205]

The New Hampshire Revolutionary War Rolls list him as having served as a private in Capt. Samuel Young's Company, Bedel's Regiment, from December 15, 1777 to March 1778, making him another Revolutionary War soldier to migrate to Greensboro. Josiah was then of Haverhill, New Hampshire. The men in this company were all from Haverhill, Lyme, Piermont, and Bath, New Hampshire. He died October 21, 1839 at Peacham, Vermont.

#12 Peleg^5 Hill (1757–1831) son of Peleg^4 and his wife Dorcas, came to Greensboro with his father and his brother James in 1791 from Saybrook, Connecticut,[206] one of the sixty-seven original proprietors who stayed in town (Figure 7.66). However, they are not listed in the 1790 Vermont census, which was actually taken in 1791. They built their cabin at the north end of the lake on the North Shore Road, near what is now known as Campbell's Corner, where they occupied two of their 100-acre contiguous lots # 4 in range 6 and lot # 4 in range 7. His third lot was #9 in range 15. Peleg^5 was a Revolutionary War soldier having served in the Connecticut Continental Line, enlisting November 1775 into the company commanded by Capt. Abraham Waterhouse.

Peleg^5 died June 23, 1831, at the age of seventy-four in Greensboro, and is buried in Noyes Cemetery. His wife was Mary Sloane (1763–1846). The Orleans County General List records Peleg Hill's holdings in 1815 as having only 100 acres of land and one house valued at $610.00. Peleg's inventory of real and personal estate was revealed in the estate papers filed September 17, 1831, by Mary Hill, executrix, and Willard Lincoln and Stephen Sherman, appraisers. The farm consisted of 130 acres valued at $800.00; hay, corn, oats, potatoes; two oxen and two three-year-old steers, two cows and two calves, a hog and two pigs, but only seven sheep, fewer than his neighbor, Willard Lincoln, whose flock numbered seventy-four sheep in 1839. There are the usual farm tools and equipment, "jointer and fore plain," three post augers, an umbrella, books, furniture, three beds and bedding, clothing, and a wooden clock. An additional appraisal on April 5, 1831, added one small house built by Samuel Hill and standing on the farm of the late Peleg Hill valued at $75.00, and one share in the Greensboro library at $2.50. Total value of his estate was $1,247.25.[207]

The following people and their properties continued up Cook Hill Road from the Bayley-Hazen Road from #13 around to #8 (Figure 7.64).

---

[205] Weber 1990:25.
[206] Stone 1854:11.
[207] Orleans County Probate Court, Vol. 2: 325.

#13 Horace Shepard moved in 1790 onto what was the Capt. Jonathan Ring property. (See Part 1.)

#14 The Ring farm, Jonathan and David Ring, on Cook Hill Road. Jonathan Ring the elder was born sometime in 1736 and died 1815 in Haverhill, New Hampshire. According to Rev. Stone, he arrived in Greensboro in 1790, although he is not in the 1790 Vermont census. His first wife was Martha (maiden name unknown); his second wife was Zilpha Adams, whom he married in July 1770. Jonathan was an early settler to Haverhill, New Hampshire, across the Connecticut River from Coos/Newbury, Vermont. He served as a sergeant and lieutenant in the New Hampshire militia for the defense of the northern frontier during the Revolutionary War. By occupation he was a carpenter and builder. His children were all born in Haverhill, New Hampshire: Sarah, born 1768, who married Joseph Ladd in 1786; Martha, born 1769, who married Horace Shepard of Newbury on March 10, 1787 (later, they lived in Greensboro and East Hardwick); and Elizabeth, born 1771, who married John Montgomery in 1789. Jonathan and his second wife, Zilpha (Adams), had six children including Jonathan Jr. born 1775, and David, born 1779. Both resided in Greensboro.

The marriage of Horace Shepard to Martha Ring explains how the Shepards came to live on the Ring Farm. A deed dated 1831[208] describes the property as being "a tract of land called the Ring Farm containing 105 acres being the whole of Lot #5 in 2nd Range, of the original proprietor Joseph Chester, and 5 acres in Lot 5–3rd Range, the original proprietorship of Simon Pettee, being the same land conveyed to Ellen D. and Charlotte Manchan by Samuel Batchelder." In another deed dated that same year,[209] David Ring grants to his son, Washington A. Ring, a dwelling house and blacksmith shop near the Burying Ground opposite Capt. Crain's house for $120.00, which reveals the probable occupation of David Ring.

Jonathan Jr. married Polly Bliss on December 28, 1797, in Newbury, Vermont. He appears in the 1800 and 1810 censuses of Greensboro after which he is said to have lived in Lisbon, New Hampshire. He may have died August 26, 1847, at Hemmingford, Canada East (Province of Quebec). Jonathan Jr. served in the 45th New Hampshire Regiment of Capt. James Flanders' unit enlisting March 21, 1814, at age thirty-six.[210]

His brother David, born 1778, married Jemimah Chamberlain April 1804 at Hardwick. At that time, he was a resident of Greensboro. David died in Greensboro April 23, 1841, at age sixty-three, and was listed in the 1815 Orleans County General List as possessing only four acres with one house

---

208 Land Records, Vol. A:454.
209 Land Records, Vol. A:35.
210 Ancestry.com Military Records, U.S. Army, Register of Enlistments 1798–1914.

valued at $100.00, while Jonathan Jr. was listed with 100 acres of land and one house valued at $600.00.

At a Probate Court held at Irasburgh June 3, 1841, Jemima Ring appears and represents she is the widow of David Ring, late of Greensboro, died intestate, and requests the Administration be granted to her. She is to make payment of the debts and settle his accounts. Elijah Sherman and Samuel Hill were appointed appraisers and asked to make inventory within thirty days. This inventory amounted to $143.88. "Since the inventory does not exceed $150.00 it is decreed that all the said estate is necessary for support of the Widow and children of the deceased forever."[211]

#15 Charles Cook Sr.

Rev. Stone claims that Charles Cook Sr. relocated to Greensboro from Danville in 1796. Yet he is enumerated in the 1800 Vermont census at Danville, so the more likely date for his appearance in Greensboro is 1801. Stone also writes, "John Law who came before July 1793 having sometime previously removed from his original pitch [near Levi Stevens?], was succeeded by Charles Cook on the farm now owned and occupied by his son Charles Cook, Esq."[212] (Figure 7.66)

Charles Cook Sr. was born September 14, 1768, at Newburyport, Essex, Massachusetts. By the time of the 1790 census, his parents, Samuel and Judith Cook and family were residing at Campton, Grafton County, New Hampshire. Charles married Elizabeth Burbeck in October 1797. Charles was about twenty-eight years old in 1796 when he moved to Danville, Vermont. They had fourteen children, some of whom died young. Their oldest child, Jane, married Thomas Tolman Jr. Elizabeth, born 1794, married Willard Lincoln; Charles B. Cook married Caroline Huntington; Caroline, born 1806, married Stephen Sherman, who frequently served as an appraiser for the estates of deceased friends.

Charles Sr. died intestate May 26, 1841, at age seventy-two years, in Greensboro. Elijah Sherman was appointed administrator with Joseph Glidden, Samuel Hill, and Enoch Tolman appointed appraisers. His estate inventory shows that he was more prosperous than some of his neighbors. No real estate is listed because it was all conveyed to his son, Charles Cook Jr., before the decease of his father. There were only six surviving heirs, each of whom received one-sixth of the distribution of $2,924.30: to Cyrus and Nancy Cook Barber, to Judith Cook Morse, to Elizabeth Cook Lincoln, to Caroline Cook Sherman, to George H. Cook, and to Jane Cook Tolman, "all being of full age, etc." The estate was discharged July 12,

[211] Orleans County Probate Court Vol. 5:233.
[212] Stone 1854:13; 1877:213.

1843, "Charles Cook, one of the heirs having received his share in said estate by real estate of which he became possessed before his father's decease."[213]

The inventory contains many notes of loans, several to his son Charles, but the listing shows items in his personal estate that inventories of the men in this study did not show: four "buffaloes," unusual for Vermont; a secretary at $20.00, the most valuable piece of furniture he owned; two large gilt pitchers; fire dogs; fifteen books including a large and a small Bible, yet he did not own stock in the library society. Charles Cook held one pew in the meetinghouse at $35.00, and two other pews at $50.00. He had "claims on the Lincoln farm and on the T. Tolman farm for a total sum of $999.90, which is doubtful." Total estate value "including uncollected demands, good, bad, and doubtful": $2,924.30.[214] Charles Cook Sr. and his wife Elizabeth are both buried in the Lincoln-Noyes Cemetery.

#16 **David Vance** first began a little north of where Mr. John C. Ellsworth now lives writes Rev. Stone in 1854, but "soon succeeded Mr. Silas Davidson where Mr. Thomas Smith now resides" on what is now Edsall Road. David Vance was not listed in the 1790 census, but was residing in Greensboro before July 25, 1793, since he was among the wedding guests (which included the whole town) who attended the first marriage in Greensboro. He was born between 1760 and 1765, possibly in New Hampshire, since he is listed in the 1790 census along with a John and James Vance at Fishersfield/Peterborough, Hillsborough County, New Hampshire. He died December 12, 1810, in Greensboro at age forty-five. He and his wife Mary were buried in the Noyes Cemetery. She died May 10, 1808, at Greensboro. A John Vance is listed as residing in Greensboro in the 1810 census.

Records are few for David Vance, except for the 1800 and 1810 censuses at Greensboro, and two entries for his death in Vermont Vital Records: one lists his death as December 19, 1810, at age fifty, birthplace not named, and the second as December 21, 1810, at the age of forty-five; burial is at the Lincoln-Noyes Cemetery. Mary, wife of David, died May 10, 1808, age forty-two.

The only Vance listed in the Orleans County General list of 1815 is James with 89 acres of land and one house valued at $800.00. One can speculate that this could be James Vance, the son of David and Mary, who died September 20, 1818, at age thirty, and is buried in Noyes Cemetery, or possibly James, the brother of David.

---

[213] Orleans County Probate Court 5:234.
[214] Orleans County Probate Court Vol. 5:233–239.

David's estate was entered for probate at Craftsbury on January 7, 1811, by the Honorable Royal Corbin, Judge. Real estate was described as the westerly half of Lots #4 in the 2nd and 3rd ranges with buildings for a value of $1,166.00. This record confirms the location of David Vance on the Edsall Road area. Also, the easterly half of Lot #3 in the 2nd range, and 5 acres of cedar land in Hardwick, value $17.00. The personal estate lists farm animals, including nine sheep, farm tools and equipment, a sleigh, kitchen utensils and furniture, the most valuable item being a cherry-wood table valued at $3.00. There was a loom with great and little wheels for spinning. With clothing, bedding and beds, and three-year-old steers, the total value of the estate was $1,906.93. But with claims against the estate taken July 7, 1812, the executor, Amos Blanchard, declared the estate insolvent. Children named in the will are James, Hiram, Orrel, Burrell, Sophia, Minerva, and Myra.[215]

#17 Amos Smith Sr. is said to have located in this area about 1792 as calculated from his Revolutionary War Pension File #S22518 of his service with Connecticut Continental Line troops. He declared in August 1832 that he was seventy-five years of age and had resided in Greensboro the past forty or so years. He was born in 1757 in Lyme, Connecticut. His death date varies from 1843 to 1845, and he claimed he was one month short of his eighty-seventh birthday. He was buried in Lincoln-Noyes Cemetery.

Amos Sr. and his wife Lucinda (Miller) had two known children, Amos Jr., who married Hannah Cate/Kate in 1809 at Walden, Vermont, and Alice, who married Jason Shepard.

Rev. Stone's history states that "Amos Sr. arrived in Greensboro before 1795 and lived where Mr. Amos Kate now lives" (in 1854). This location appears on the 1859 H. W. Walling map as ""Amos Smith Jr., being husband of Hannah Kate, located uphill from Charles Cook and the start of Edsall Road."

#18 Silas Davidson was said by Rev. Stone to have been settled in Greensboro in 1793[216] on what is now Edsall Road. Little information is available about this man. He was not listed in the 1790 Vermont census (actually taken in 1791), but he and his wife were among the wedding guests at Joseph Stanley's marriage to Mary Gerould on July 25, 1793. He does not appear in the 1800 census for Greensboro. Is Silas related to Paul Davidson listed in Greensboro in the 1790 census? A Paul Davidson is enumerated in Brookfield, Vermont, in 1800.

---

[215] Orleans County Probate Court, Vol. 1:105, 187.
[216] Stone 1854:212.

Comparing the 1800 Greensboro census with the present neighborhood study, starting with Levi Stevens who lived about where the John Allen house stood across from the old Hazendale Farmstand, extending up to Campbell's Corner at the north end of the lake, the order in which the census was enumerated closely matches the order of homeowners mentioned on the interpretive map (Figure 7.64 above, and Weber 1990:24). The only exceptions were Aaron Farnham and William Moore, who were listed in the 1810 census as being situated between Jonathan Ring, Jr. up Cook Hill, and Horace Shepard on what was the Ring farm.

Additionally, an obscure manuscript was shown to this writer on a visit to see David Linck at the Craftsbury Historical Society June 23, 2010: the water-stained and tattered Orleans County General List of 1815, which contained a listing of all Orleans County towns with resident and non-resident property owners of that year. The manuscript was unsigned but was probably authored by Samuel Crafts. This manuscript has not yet been photocopied or deposited at the Vermont State Archives at the time of this writing. The General List included the names of the property owners for all Orleans County towns, as well as the number of houses and outbuildings, their dollar value, and another column termed "equalized," which contained a lower value than the dollar value of each property.

Summary

The military importance of the Bayley-Hazen Road diminished, but the fact remains that if this road had not been constructed, the early settlement of the region may have taken much longer to establish, or quite possibly, this wilderness region might not have been settled at all. However, once settled, this became a productive region, not only to those living in it, but also for trade, personal reasons, and expedient passage to southern Canada from areas such as Boston.

The pioneers and their families who populated the early Greensboro neighborhoods laid the foundations of the town, literally and figuratively. Thanks to the efforts of these early settlers, residents of Greensboro today enjoy a vibrant community in a unique and beautiful setting. To learn more about the additional settlers in other parts of town, we invite you to read Patricia Haslam's article on this topic in the *Hazen Road Dispatch* (2014).

Chapter 8

Preliminary Interpretation and Conclusion

The purpose of this work is to share and make available the data we have extracted from the site we consider to be that of the Greensboro, Vermont, Revolutionary War-period blockhouse. After only four seasons of limited excavation, it would be premature and somewhat speculative to offer any definitive conclusions about this site. However, it is appropriate to offer some preliminary observations and suggestions while acknowledging that data produced from future excavation may challenge and critically revise those preliminary conclusions presented here, albeit with a degree of confidence.

According to Elkins (1921), Hazen arrived in Peacham in 1779 with approximately 150 men in his regiment. They cut through the forest to construct the road and bridges, and they built blockhouses, one of which was in Greensboro, roughly 18 to 20 miles north of Peacham. As has been stated, the original purpose of the blockhouse was militaristic in nature and intended to house soldiers. Based on Elkins' description, the blockhouses were constructed with hewn timbers and fieldstone and contained port holes for the use of small arms. Whether any other materials, such as nails or bricks, were transported to the site for construction remains unknown. In his reference to the soldiers who had been sent northward to scout near the Greensboro blockhouse, some of whom were killed or kidnapped from the site in 1781, Elkins mentions that they had stopped there to "cook some victuals" (1921:206). This could imply there was at least a fireplace, and perhaps the fort was regularly outfitted with supplies such as firewood, flints, and pots and pans for cooking utilized by soldiers who stayed at the blockhouse temporarily or were stationed there.

Based on the 15 x 15-meter grid established in 2010, over the course of four seasons not more than 32.5 square meters were excavated, approximately 14.44% of the grid has been investigated to date. According to Elkins' (1921) description of the Peacham, Walden, and Greensboro blockhouses, it is probable that a stockade fence surrounded the Greensboro blockhouse. It is, therefore, possible that the site extends some 15 meters (49.50 feet/16.5 yards) beyond the walls of the structure we have identified, and remains unexplored.

The portion of the foundation and outer walls we excavated were substantial. Measuring 2 to 3 meters wide (stone foundation + rampart) in Probe 2 (2012), the foundation and wall excavated there were made of mostly unworked fieldstone. The foundation trench was dug well into the ground and filled with fieldstone, providing a solid footing for the superstructure. Based on the bottom level of the 2015 season, it appears that the foundation trench in Probe 2 was well below the lowest level of the hearth/chimney platform/joist (Loci 2/3). The outer wall of the structure was built on that foundation, with an earthen berm, or rampart, that extended outward (northward) from the wall. This was made of a light brown hard-packed clay-like soil, which was very different from the dark black soil found in Probe 1 and Square NE4. The berm also rose well above the surrounding forest floor thus elevating the overall height of the wall. This suggests that the soil comprising the rampart was purposefully deposited there. A berm such as this not only added strength and protection to the lower part of the wall but also probably served as a defensive mechanism meant to prevent sapping or collapsing of the structure by would-be intruders. The substantial nature of the foundation, wall remnant, and earthen berm suggest this structure was intended for use as a fort rather than a farmhouse.

Inside the structure, south of the foundation wall, was a thick hard-packed layer of soil and crushed shale, dark gray in color. This layer extended into the balk, so the full extent of this layer is not yet known. Since this layer was located directly below the stone foundation/wall, it may have been a deliberate fill lining, the foundation trench, which would have provided a substantial base for the foundation. If, however, this crushed shale layer extends throughout the interior of the structure, then perhaps this was the original floor. A thick layer of crushed shale would have provided solid flooring and would have been useful if those stationed at the fort needed to shelter horses or other livestock, as it would have absorbed liquids that may have spilled, and would have been relatively fireproof.

Found among the collapsed portion of the structure in Probe 2 were numerous worked stones. These stones appear to have had right angles cut into them and probably functioned as architectural elements that would have seated timber beams for the superstructure and/or flooring.

Based on the number, type, and location of nails found throughout Probe 1 and Sq. NE4 and the joists, it can be said with some certainty that there was a timber floor. What is not certain is whether the wooden floor was part of the original construction or added later, either by the military or by the civilian inhabitants. No nails were discovered in Probe 2 and most of the nails that were found were located in the center of the building. This may suggest that the main structure of the blockhouse was constructed using wooden pegs and joints such as mortise and tenon and dovetail, while the interior finishing required nails, which would confirm Bogart's description of the blockhouses being "... pinned at the corners and perhaps at intervals along their length".[217] Even though no wood was recovered during excavation, the nails and modified stone architectural elements attest to the presence of wood having been there at one time.

Throughout Square NE4 and to a lesser extent in Probes 1 and 2, bricks were found. The greatest concentration of them appears to have been in and around Sq. NE4. Presumably the hearth, firebox, and chimneystack were made primarily from fieldstone. However, the presence of brick, both whole and halved, suggests that at least part of those components may have been made of brick. Alternatively, it could be that the original hearth, firebox, and chimney were constructed of fieldstone and later modifications to the firebox, or the addition of an oven, may have been constructed using brick. Curiously, none of the bricks contained the remnant of mortar, and no remnant of mortar was found during excavation. The bricks recovered during excavation were not enough to have constructed an entire fireplace and chimney.

It was not uncommon for blockhouses to be repurposed as residential dwellings or schools once their original military purpose had ended, which is what happened with the Greensboro blockhouse. The Fort Pitt blockhouse in Pittsburgh, Pennsylvania, has a similar habitation sequence, functioning first as a military fort and later as a residential dwelling. In 2003 the floor of the Fort Pitt blockhouse was excavated and documented so that a modern floor and heating system could be installed. The excavators, Cultural Resources Section of Michael Baker Jr., Inc., recovered over 6,700 artifacts including glass, metal, ceramic, and other miscellaneous artifacts. This excavation also revealed that the blockhouse had been modified by later civilian occupants. Originally that blockhouse included a fireplace dating to the 19th century, which was when the blockhouse went out of use as a fort and was used as a residence. In 1894 the fireplace was removed and a staircase built in its place, allowing visitors to access the second floor.[218]

The original flooring was also discovered at the Fort Pitt blockhouse. The floor was made of timber, and although it was soggy and somewhat

---

[217] Bogart 1948:49.
[218] Fort Pitt Blockhouse http://www.fortpittblockhouse.com/archeology/.

decayed, it remained *in situ*. The planked timber flooring was covered in the mid-1800s when brick was placed over it as part of the modifications made during the residential phase. In the middle of the floor a trench measuring 6.4 feet deep has been interpreted as a storage bin used from ca. 1785–1894 during the residential phase.[219] Based on this example, it is possible that modifications were made to the Greensboro blockhouse after the Shepards took ownership of it. For example, if the original floor of the blockhouse had been hard-packed dirt/clay or crushed shale (or both), the residents may have retrofitted the structure with accouterments such as timber floors. By inserting worked stone in specific locations and adding stone joists next to the hearth/chimney platform and spanning the voids between the structure's walls, timber beams and floors could have been added. The space between the ground and the planks would have provided a barrier against the cold and moisture of the original ground floor. Similarly, modifications could have been made to the fireplace with the addition of ovens for more nuanced cooking suitable to civilian life.

After the death of Aaron Shepard and the sale of the blockhouse to Joel Cushing, it may be that the blockhouse exhibited signs of collapse. Curiously, no wood, no negative impression of disintegrated wood, nor any staining from rotted wood has been found during these four seasons of investigation. Nor has there been any evidence of burning other than a few small bits of charcoal and an ash pit found in the northwest corner of Square NE4, which is likely related to the fireplace where cooking was conducted. Furthermore, the stone walls "collapsed" inward toward the center of the building rather than randomly or in a twisted manner as usually happens when a building collapses, which leads to the conclusion that the blockhouse was deliberately demolished. If the subsequent owners knew the building was on the verge of collapse, perhaps they salvaged as many usable materials as possible, such as wood, nails, and bricks, and used them to build a new structure nearby.

Similarly, thus far, no military paraphernalia, such as musket balls, flints, coins, buttons, etc. have been recovered. When excavating a structure with known military purposes, one would expect to find military-related artifacts, so it may seem surprising that such items were not recovered. However, our excavation activities took place inside the structure of the blockhouse, not outside its walls, which is where one would expect to find spent ammunition. Additionally, since the blockhouse was so infrequently inhabited by military personnel, it is not likely they would have left behind large quantities of military-related items. Furthermore, the subsequent civilian residential phase may have obliterated much of the military material culture if that had been left behind, which is similar to the Fort Pitt habitation sequence.

---

[219] Fort Pitt Blockhouse http://www.fortpittblockhouse.com/archeology/.

The majority of the artifacts recovered through archaeological excavation from this site represent the residential phase rather than the military one.[220] Therefore, the fact that excavations at the Greensboro blockhouse site have thus far produced more artifacts from the residential phase than the military occupation is not without precedent and does not contradict or disprove the structure's original purpose.

The pottery found at the site can be classified as mostly utilitarian, storage, and service ware consisting of tableware. These ceramic types were commensurate with most contemporaneous New England and American households and commonly found in the ceramic market.[221] Tools such as the hoes, a whisk, and a pair of scissors may be indicative of domestic activities such as farming, sewing, and, of course, food preparation, which is underscored by the animal bones. Together, these may also indicate a comfortable to somewhat elevated socio-economic status.

The importance of this site, the associated military road, and the blockhouses has been questioned by some and celebrated by others. One way to measure the importance of a practice, campaign, or project is to consider the energy expenditure hypothesis, which takes into consideration the amount of energy that was spent during that practice or project. For example, in the building and preparation of a tomb for a deceased individual, the amount of energy spent on the preparation and size, as well as the complexity and equipping of an individual's tomb, may be commensurate with that person's rank. Therefore, deceased individuals with large, elaborate, and well-equipped tombs may be considered to be of high social status and much loved among family and the community because of the amount of energy that went into tomb preparation.[222] Similarly, the amount of energy that went into the planning, financing, and building of the Bayley-Hazen Military Road and the four blockhouses on the part of Generals Washington and Bayley, Colonel Hazen, and the others who worked on the project, must be taken into consideration when assessing whether this project had value or merit. Based on correspondence between George Washington, Jacob Bayley, and Moses Hazen discussing "…cutting a Road from Coos to St. Johns…"[223] compensation for labor and reimbursement for materials, and Congress' "… strong desire to undertake an Expedition against Canada …"[224] it would appear that this venture was of great importance initially.

---

[220] Fort Pitt Blockhouse http://www.fortpittblockhouse.com/archeology/.
[221] Diagnostic Artifacts in Maryland, http://www.jefpat.org/diagnostic/ColonialCeramics/Colonial%20Ware%20Descriptions/Creamware.html.
[222] Wason 1994:76–78.
[223] Washington to Bayley, October 17, 1776. Washington Papers http://memory.loc.gov/cgibin/query/r?ammem/mgw:@field%28DOCID+@lit%28gw060159%29%29
[224] Washington Papers.

Whether this road was meant to be a functional military artery, a ruse to fool the British, or a project pushed through Congress by greedy speculators (Bayley and Hazen), its significance cannot be underestimated. Although its military purpose was short lived, the Bayley-Hazen Military Road paved the way for the establishment of new communities, such as Peacham, Walden, Greensboro, and Cabot in Northern Vermont. This episode may also have contributed to the establishment of the Canadian border, helping those Revolutionaries to realize a logical place to locate the northern extent of the emerging United States.

---

http://memory.loc.gov/cgibin/query/r?ammem/mgw:@field%28DOCID+@lit%28gw130269%29%29.

Appendix 1:
Fisk Letter (1903)
Greensboro Association Report (1941)

Appendix 2:
Master Grid Plan

List of Maps and Records

References

Index

Appendix 1

Fisk Letter (1903)

In 1903 Perrin B. Fisk discussed with Brother Cook the erection of a granite stele in honor of the two soldiers, Moses Sleeper and Constant Bliss, who were killed while occupying Greensboro's blockhouse. Shown below is page one of the letter.

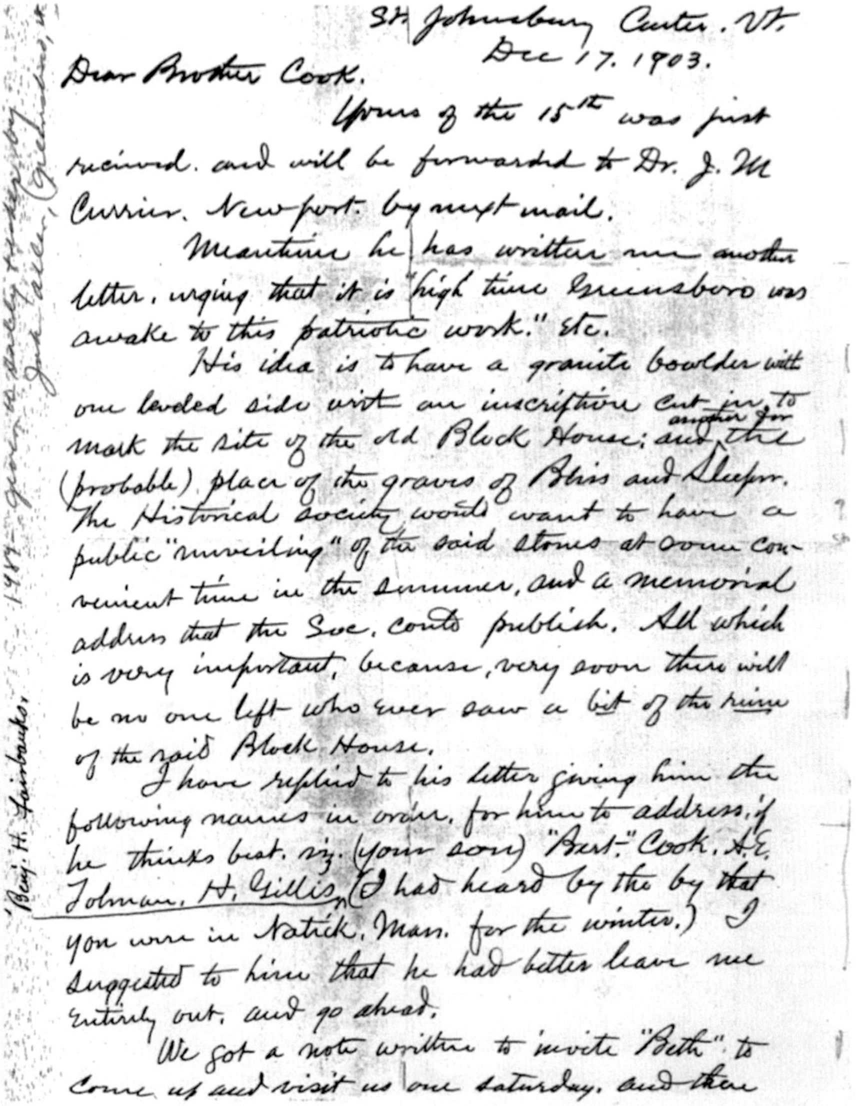

St. Johnsbury Center. Vt.
Dec 17. 1903.

Dear Brother Cook.

Yours of the 15th was just received. and will be forwarded to Dr. J. M Currier. Newport. by next mail.

Meantime he has written me another letter. urging that it is "high time Greensboro was awake to this patriotic work." Etc.

His idea is to have a granite boulder with one leveled side with an inscription cut in, to mark the site of the old Block House; and another for the (probable) place of the graves of Bliss and Sleeper. The Historical Society would want to have a public "unveiling" of the said stones at some convenient time in the summer, and a memorial address that the Soc. could publish. All which is very important, because, very soon there will be no one left who ever saw a bit of the ruins of the said Block House.

I have replied to his letter giving him the following names in order. for him to address, if he thinks best. viz. (your son) "Bert." Cook. A.E. Tolman, H. Gillis, (I had heard by the by that you were in Natick. Mass. for the winter.) I suggested to him that he had better leave me entirely out. and go ahead.

We got a note written to invite "Beth" to come up and visit us one saturday. and there

Benj. H. Fairbanks,

The letter also discusses establishing a stele to mark the location of the blockhouse. Shown below is page two of the letter.

learned that she had gone home, sick. Since
then, for one reason and another, we have
not seen a good time for her to come. If
she comes back next time, we will endeavor
to secure to ourselves that pleasure, provided
she is willing to ride up with me, and visit
two rather lonely people. We would be glad
to entertain any of your family at any time.
Fidelia has been sick (in Boston) during Thanks-
giving week, and had to do double work before she
was fit to be back in the office, as her assist-
ant fell sick before she was well. Grace has
the care of a minister's mother in Peterboro,
N.H. The poor woman came there on a
visit; had a shock, and never can go home, —
May live a week or years. Is generally a very
pleasant case, but has severe & protracted
crying spells, which must be very trying.
Flora I visited last week @ Lunenburg, to
give the Woman's Club (at her house) an address
on "Lanier". She & Nina were well as usual.
Nina is at school (and doing well) till 2:30 P.M.
and then takes care of the Telephone Central,
which is in her mother's house, and care.
Mrs. Fisk sends ... to Mrs. Cook, and I, greet-
ings & regards to all your circle. Perrin B. Fisk

(over)

Greensboro Association Report (1941)

The following image is the 1942 report of the Blockhouse Memorial Committee to the Greensboro Association, which discussed the details of having erected the granite stele.

Greensboro Vermont
August 15 1942.

Report of

THE BLOCKHOUSE MEMORIAL COMMITTE
TO THE
Greensboro Association:

The Fletcher granite monolith was set on the foundation September 13' 1941 and the grading and seeding was completed September 24" 1941.

Due to erosion the plot of land now needs grading and seeding in some spots and this is having the attention of the committee. The committee is also discussing and inquiring as to the desirability of planting some native shrubs and trees- possibly cedars* at the boundry of the lot. However it is realized that nothing of the sort should be done that will in the future obstruct the beautiful view of the Lake.

The foundation is five feet deep, resting on the ledge of rock which underlies the area surrounding the location of the site. It is reinforced with iron rods placed vertically around monument which extends eighteen inches intothe cement.

The Committe verified the historical accuracy of the names on the monument by inquiring of the Vermont Historical Society and of Dr. Bliss Perry who has access to the Bliss family records. Letters from these two sources are attached and are a part of this report as exhibits "A" and "B" respectively.

The Historical Society letter states in part as follows- "Th names of the two killed at the Block House in Greensboro in 1781 are Constant Bliss of Thetford and Moses Sleeper of Newbury. Their names are to be found on p.p. 72-73 of Miller and Wells historyof Ryegate published in 1913". Mr. Charles Bliss in a letter to Dr. Bliss Perry stats that the Bliss Family records contain the following item "Constant A. Bliss 1746 ,, scalped by the Indians at Greensboro Vermont."

Dr. Bliss Berry drafted the inscription.

To date seventy five individuals have made cash contributions totaling $203.75 and The Orleans County Historical Society contributed $10.00 making total cash received $213.75
The total cost is as follows:

Monument lettered and erected .	$106.45
Foundation ,, labor and material	14.13
Grading and seeding	30.00
Postage telephone and sundries . . .	14.20
Total cost	$164.73

Excess cash receipts over expenses $49.02.

Inaddition to the above expenses The Selectmen gave one days labor of four men and the use of their road truck, cleäring brush and levelling the site. This represents a value of not less than $20.00 Another significant gift came throuhh the generosity of Mr. and Mrs. Wilfred Mercier whogave the plot of land on which the monument stands.This plot is abóut forty by sixty feet in size, extending from the center of the highway to a line eleven feet Easterly of the rear surface of the monument proper.The N.E. and S.E.cornes of this plot are marked by granite markers set in the ground.

The Committee assumes that this fund on hand in excess of the cost will be retained to be drawn upon from time to time for the care and upkeep of the monument and site as may be needed.

Title to the land is in the name of The Town of Greensboro.

The Committee is gratified with the cordial reception accorded their efforts in securing this fund and is especially happy that the project represents the interest of many people rather than that it should have been the gift of a few.

It is greatly to the crédit of our citizens --both full time and Summer time residents- that one of the most interesting historical sites in the state of Vermont is now suitably marked and by the most appropriate and enduring matrial, VERMONT GRANITE..

By the Committee:.

George C.Hubert Chairman

J. H. Barrington

Mrs. Donald Frazer

Mrs. John Minor

Miss Alice Snyder

Appendix 2
Master Grid Plan

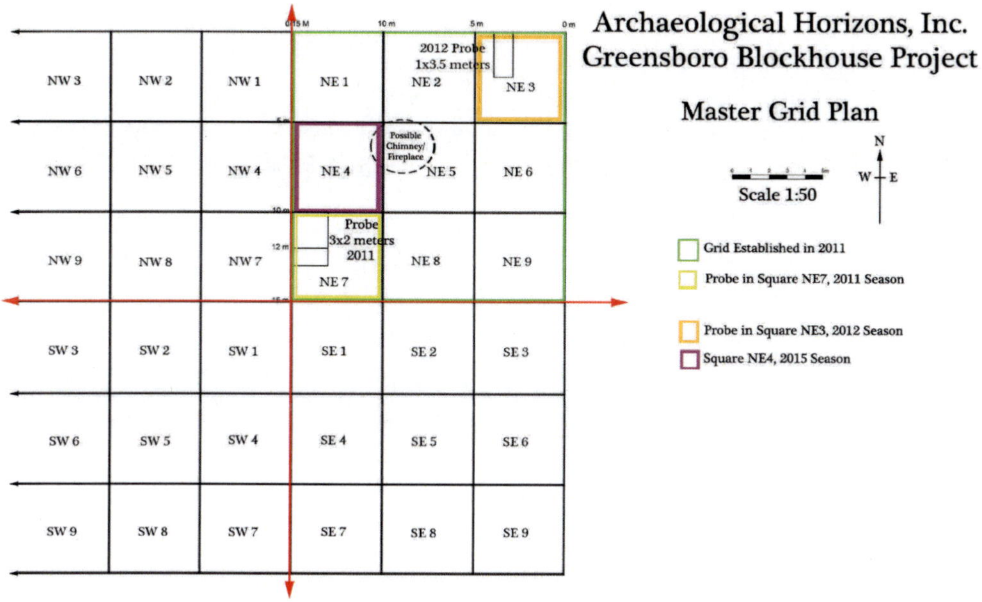

Master Grid Plan. The artificial grid that was superimposed over the site for the purpose of controlled excavation and recording. The Surface Survey of 2010 was conducted within the northeastern quadrant (Squares NE1-9). Probe 1 (2011), Probe 2 (2012), and Square NE4 were excavated and are indicated on the plan. (Plan prepared by Jill L. Baker.)

List of Maps and Records

Maps

Abel Sawyer map, 1784 [Cobb entry #53]

Blodgett/Amos Doolittle map 1789 [Cobb#68]

Jonathan Elkins pocket map [Peacham Town Clerk[

James Whitelaw 1790 [Cobb #84]

Carey 1795 [Cobb #118]

J. Whitelaw 1796 [Cobb #122]

Sotzman/Bohn 1796 [Cobb #124]

Records

Cemetery Census, Greensboro, VT. Compiled July 1980. Manuscript found at the Greensboro Historical Society.

Greensboro Historical Society Archives, genealogies.

Land Records, Town Records, Cemetery Records, Vital Records. Found at the Greensboro Town Clerk's Office, Greensboro, VT.

Orleans County General (Grand) List. 1815 Manuscript found at the Craftsbury Historical Society, Craftsbury, VT.

Orleans County Probate Records. Vermont State Archives & Record Administration. Microfilm.

U.S. Revolutionary War Rolls 1775–1783, National Archives Microfilm Rolls, M246, Record Group 93, Washington, DC.

References

Ancient Discoveries. 2008. *Ancient Mining Machines.* Season 5, Episode 5. 20 November 2008. Accessed 23 June 2017.

Archaeology and Anthropology Collections. (2015). Staffordshire Pottery Excavated in Middletown. Wesleyan University. Retrieved from http://www.wesleyan.edu/archprog/collections/collection-highlights/staffordshire.html. Accessed 2 September 2015.

Baldwin, F. W. (1906). History of the Hazen Military Road. *The Vermonter,* Vol. XI No. 16. Pp. 297–323. (Retrieved from: http://www.rootsweb.ancestry.com/~vermont/HistoryHazenMilitary Road.html).

Bailey, F. (1926). Colonel Frye Bailey's Reminiscences. *Proceedings of the Vermont Historical Society, for the Years 1923, 1924, and 1925.* Bellows Falls, VT: The P. H. Gobie Press, Inc.

Beers, F. W. (1875). *County Atlas of Caledonia, Vermont. From Actual Surveys by and Under the Direction of F. W. Beers.* New York: F. W. Beers & Co.

Bentley, P. K. (1999). *Ancient Siege Warfare.* Bloomington and Indianapolis: Indiana University Press.

Bogart, E. L. (1948). *Peacham, The Story of a Vermont Hill Town,* Montpelier, Vermont Historical Society. Pp. 40, 42. (Reprinted 1981 by The Peacham Historical Association. University Press of America, Inc.).

Bogart, E. (1981). *Peacham, The Story of a Vermont Hill Town.* (First published 1948). The Peacham Historical Association. University Press of America.

Bonn, C. (1796). Map of Vermont. Retrieved from http://www.old-maps.com/vt 1796 Stotzmann Im.jpg.

Bourriau, J. D., Nicholson, P. T., Rose P. J.. (2006). Pottery. In P. T. Nicholson and I. Shaw (eds.) *Ancient Egyptian Materials and Technology.* Cambridge University Press. Pp. 121–147.

Brinton, C. (1955). The Hazen Road. *Vermont Life Magazine.* XI, 3:32–38.

Brown, C. (1894). *History of the Shepard Family.* Montpelier: Press of the Argus & Patriot Co. Accessed through GoogleBooks.com (30 July 2012). Also, E. Hardwick, VT: Vermont Historical Society, Barre, VT.

Child, H. (1887). *Essex and Caledonia Counties Gazetteer, 1764–1887.* Syracuse, NY: The Syracuse Journal Company, Printers and Binders.

Child, H. (1889). *Gazetteer of Washington County, VT. 1783–1889. First Part.* Syracuse, NY: The Syracuse Journal Company, Printers and Binders.

CastlesFortsBattles. (2016). Inchtuthil Roman Fortress. Retrieved from http://www.castlesfortsbattles.co.uk/inchtuthil_roman_fort.html. Accessed 18 January 2016.

Cobb, D. A. (1971). *Vermont History. The Proceedings of the Vermont Historical Society, Special Double Issue, Vermont Maps Prior to 1900: An Annotated Cartobibliography.* Vermont Historical Society, Vol. XXXIX, No.'s 3 & 4, (Summer & Fall 1971) in which "forts" were described on maps before 1810.

Correspondence.
Between Giovanna Peebles, State Archaeologist, Agency of Development & Community Affairs, Montpelier, VT., and Pat Haslam, 31 Oct. 1991.

Correspondence.
Between Giovanna Peebles, State Archaeologist and Francis Foster, RFD 2 (Walden, VT), Hardwick, VT, she requesting permission to visit the site, and enclosing a copy of the site survey form (1985). [The site visit never happened.]

Crockett, W. H. (1921). V*ermont The Green Mountain State.* New York: The Century History Company.

Cummings, A. L. (2003). *The Fairbanks House. A History of the Oldest Timber-Frame Building in New England.* Boston: New England Historic and Genealogical Society.

Dictionary of Fortification, Civil War Field Fortifications Website. Blockhouse. Retrieved from http://lly.org/~rcw/cwf/dictionary/xgb-017.html. Accessed 2 November 2015.

Donnelly, M. C. (2003). *Architecture in Colonial America.* Eugene: University of Oregon Press.

Duffy, J., Hand S. B., Orth, R. H.. (2003). "Bayley, Jacob" and "Bayley-Hazen Military Road". P. 51 in *The Vermont Encyclopedia.* Burlington, VT: The University Press of New England.

Eddy, M., Eddy, O. (1958–1959). *The Hazen Military Road, Town of Walden.* Unpublished manuscript, held at the Walden Public Library.

Elkins, J. (1921). Reminiscences of Jonathan Elkins, [1774–1781] From a Manuscript in the Possession of the Vermont Historical Society. *Proceedings of the Vermont Historical Society for the Years 1919–1920.* Pp. 187–214.

Elkins, J. (1943). Reminiscences of Jonathan Elkins. *The Upper Connecticut,* Vermont Historical Society Collections. 1:263–289.

English, V. H., McCorison, M. A.. (1959). *Atlas of The Bayley-Hazen Road 1776 & 1779.* Lyndonville, VT: Northeastern Vermont Development Association.

Fisher, C. E., Fisher, S. G. (1992). *Soldiers, Sailors, and Patriots of the Revolutionary War, Vermont.* Rockland, ME: Picton Press.

Fiske, P.B. (1901). *Decoration Day Address at Greensboro.* Manuscript at the Vermont Historical Society. Retrieved from www.northeastamericanforts.com/East/vt/html.

Gale, R. P., Gasson, N., Killen, H. and Killen, G. (2006). Wood. In P. T. Nicholson and I. Shaw (eds.) *Ancient Egyptian Materials and Technology.* Cambridge University Press. Pp. 334–371.

Goodrich, J. E. (comp. and ed.). (1904). *Rolls of the Soldiers in the Revolutionary War 1775–1783.* Rutland, VT: The Tuttle Company. Edited and reprinted, by C. L. Knapp.

Greensboro Proprietors Map. (1918). James Whitelaw Surveyor General's Papers, Vermont State Papers, Vermont Archives and Records Administration. Vol. 1.

Guilland, H. F. (1971). *Early American Folk Pottery.* Chilton Book Company.

Hammond, I. (ed.). (1889). R*olls and Documents Relating to the Soldiers of the Revolutionary War.*

Haslam, P. (1981). Some Additions & Corrections: Soldiers of the Revolutionary War Buried in Vermont: [Washington County].

Vermont History: The Proceedings of the Vermont Historical Society. 49(4):232–245; reprinted (Fall 1984) 52(4):225–242.

Haslam, P. (1990). The Old Stone House. In S. B. Weber, ed., *The History of Greensboro: The First Two Hundred Years*. Greensboro: Greensboro Historical Society. Pp. 51–53.

Haslam, P. (1995). Aspenhurst Farm & the Tolman Connection: the 200th Anniversary of Thomas Tolman Jr. to Greensboro. *Hazen Road Dispatch*. Journal of the Greensboro Historical Society. 20:35–37.

Haslam, P. (1996). New Discovery: Following Up on the Date for the Stone House. *Hazen Road Dispatch*, Journal of the Greensboro Historical Society. 21:16–18.

Haslam, P. (1997). A Hike to the Garrison at Walden, September 1985. *Hazen Road Dispatch*. Journal of the Greensboro Historical Society. 22:6–8.

Haslam, P. (2011). Near This Spot.... The Story About the Bliss/Sleeper Monument. *Hazen Road Dispatch*. Journal of the Greensboro Historical Society. 36:9–12.

Haslam, P. (2012). Greensboro's Earliest Neighborhood, Part 1. *The Hazen Road Dispatch*. Journal of the Greensboro Historical Society. 37:13–17.

Haslam, P. (2013). Greensboro's Earliest Neighborhood, Part 2. *The Hazen Road Dispatch*. Journal of the Greensboro Historical Society. 38:4–14.

Haslam, P. (2014). Additional Settlers in Other Parts of Town. *Hazen Road Dispatch*, Journal of the Greensboro Historical Society. 39:21–30.

Hemenway, A. M. (1867). *The Vermont Historical Gazetteer. A Magazine, Embracing A history of each Town. Civil, Ecclesiastical, Biographical, and Military. Addison, Bennington, Caledonia, Chittenden and Essex Counties*. Vol. I. Miss A. M. Hemenway/ Claremont, NH: The Claremont Manufacturing Co. Accessed from Archive.org, Retrieved from https://archive.org/stream/vermonthistorica31heme#page/n3/mode/2up.

Hemenway, A. M. (1871). *The Vermont Historical Gazetteer: A Magazine Embracing A History of Each Town, Civil, Ecclesiastical, Biographical and Military. Franklin, Grand Isle, Lamoille and Orange Counties Including Also The Nautical History of*

Chittendon County and Index to Volume I, Completed. Vol. II. Burlington, VT: Published by Miss A. M. Hemenway. Retrieved from Archive.org https://archive.org/stream/vermonthistorica02heme#page/n9/mode/2up.

Hemenway, A. M. (1877). *The Vermont Historical Gazetteer: A Magazine Embracing a History of Each Town, Civil, Ecclesiastical, Biographical and Military. Orleans and Rutland Counties.* Vol. III, Part 1. Claremont, NH: The Claremont Manufacturing Company. Retrieved from Archive.org, https://archive.org/details/vermonthistorica03heme.

Hemenway, A. M. (1882). *The Vermont Historical Gazetteer: A Local History of All the Towns in the State, Civil, Educational, Biographical, Religious and Military. The Towns of Washington County, with Histories of The Town of Swanton, in Franklin County, Groton, in Caledonia County and the Additional History of the Town of Hubbardton, in Rutland County.* Vol. IV. Montpelier, VT: Vermont Watchman and the State Journal Press. Retrieved from Archive.org, https://archive.org/details/vermonthistorica04heme.

History Committee; Walden Public Library. (1986). *A History of Walden, Vermont.* Randolph, VT: Green Hills Books.

Hume, I. N. (1969). *A Guide to Artifacts of Colonial America.* University of Pennsylvania Press.

Isham, N. M. (2007). *Early American Houses with A Glossary of Colonial Architectural Terms.* Mineola, NY: Dover Publications, Inc.

Isham, N. M. and A. F. Brown. (1900). *Connecticut Houses an Historical and Architectural Study.* Providence, RI: The Preston and Rounds Company.

Johnston, H. P. (ed.). (1889). *The Record of Connecticut Men in The Revolution, 1775–1783.* Hartford, CT.

Jones, P., Derek, R. (1982). "The military effectiveness of Arrow Loops: Some experiments at White Castle", *Château Gaillard: Etudes de castellologie médiévale, Centre de Recherches Archéologiques Médiévales,* IX–X: 445–456.

Kaufmann, J. E.. Kaufmann, H. W. (2007). *Fortress America. The Forts That Defended America 1600 to the Present.* Cambridge, Ma: Da Capo Press.

Laws, B. (2014). *The History of the Garden in Fifty Tools*. University of Chicago Press.

LeFever, G. (2008) Forged and Cut Iron Nails. Early American Life. June, pp. 60–69.

Richardson, A. (2013). Archaeology Lab Ceramics Database. Saint Mary's University, Anthropology Department, Retrieved from http://www.smu.ca/academics/departments/archaeology-lab-ceramics- database.html. Accessed 29 August 2015.

Manuscript Vermont State Papers. (177–1799). Vermont State Archives and Records Administration. MVSP Vol. 18.

Morrison, H. (1952) *Early American Architecture From the First Colonial Settlements to the National Period*. New York: Oxford University Press.

McCorison, M. (1959). The Bayley-Hazen Road. *Vermont History*, Vermont Historical Society, Montpelier, VT. Vol. 27. No. 1.

Miller, E., Wells, F. P. (1913). *History of Ryegate, Vermont From its Settlement by the Scotch-American Company of Farmers to the Present Time*. St. Johnsbury, VT. Retrieved from http://www.electricscotland.com/History/ryegate/9.htm.

Mueller, J. W. Forts and Fortifications. (2017, May 8) Encyclopedia of the New American Nation. *Encyclopedia.com.* Retrieved from http://www.encyclopedia.com/history/encyclopedias-almanacs-transcripts-and-maps/forts-and-fortifications. pp. 48–52

Nelson, L. H. (1968). Nail Chronology as an Aid to Dating Old Buildings. Technical Leaflet 48. Nashville: American Association for State and Local History. Vol. 24. No. 11.

Nicholson, P. T., Shaw, I. (2009). *Ancient Egyptian Materials and Technology*. Cambridge University Press.

Ogden, J. (2006). Metals. In P. T. Nicholson and I. Shaw (eds.) *Ancient Egyptian Materials and Technology*. Cambridge University Press. Pp. 148–176.

Oleson, J. P. (ed.) (2008). *The Oxford Handbook of Engineering and Technology in the Classical World*. Oxford: Oxford Uniersity Press.

Peterson, C. E. (1950). Burning Buildings for Nails. *Journal of the Society of Architectural Historians*. Vol. 9. No. 3. P. 23. Retrieved from http://www.jstor.org/stable/987459?origin=JSTOR-pdf. Accessed: 22/09/2011 13:54.

Plumley, C. A. (1924). *Programme*. Plumley, (President, Presiding) "At the Dedication of a Marker Erected by the State of Vermont Society, Sons of the American Revolution and the Peacham Historical Society, August 13, 1924, Beside the Hazen Road, and to Mark Three Other Historic Sites at Peacham, Vermont."

Porter, A. (1984). *The China Journal of Amos Porter, 1802–1803*. Greensboro, VT: Greensboro Historical Society.

Post, B. D. (2006). *The Founding Families of Craftsbury, Vermont: A Study of the Relationships Between the Proprietors and the First Settlers of the Town*. Published by B. D. Post.

Program at Dedication of Roadside Marker. (1924). Vermont Chapter, Sons of the American Revolution, "Peacham Pioneers," 1924, p. 30.

Rolls of the Soldiers in the Revolutionary War, State of Vermont 1775–1783. 1904. Rutland, VT: John E. Goodrich, Tuttle Co.

Sangree, G. (2003). Two Letters of Thomas Tolman the Younger. *Hazen Road Dispatch*, Journal of the Greensboro Historical Society. 28:25–27.

Sprague, J. F. (1921). Fort Halifax. *Sprague's Journal of Maine History*. Vol. 9. No. 3. Pp. 132–135.

State Papers of Vermont, Vermont Secretary of State. (1918). *Index to the Papers of the Surveyors-General*. Volume 1. Rutland, VT: The Tuttle Company.

Stone, Rev. James T. (1854). *A History of Greensboro and the Congregational Church: a Semi-Centennial Discourse, delivered Nov. 24, 1854*. Service bulletin, no publisher provided.

Stone, Rev. James T. (1877). Greensboro. In A. M. Hemenway, *The Vermont Historical Gazetteer: A Magazine Embracing a History of Each Town, Civil, Ecclesiastical, Biographical and Military*. Vol. III. Part 1. Claremont: The Claremont Manufacturing Company.

Swift, E. M. (1977). *Vermont Place Names: Footprints of History*. The Stephen Greene Press, Brattleboro/ Camden, ME: Picton Press.

Thompson, Z. (1842). *History of Vermont, Natural, Civil, and Statistical, In Three Parts. Part Second Civil History of Vermont.* Burlington, VT: Chauncey Goodrich.

Thompson, Z. (1848). *Geography and Geology of Vermont.* Burlington, VT: Chauncy Goodrich.

Thorson, R. M. (2005). *Exploring Stone Walls. A Field Guide to New England's Stone Walls.* New York: Walker & Company.

Vermont Division for Historic Preservation, Montpelier, VT. (1985). Walden Site No. VT-CA-24, Field site No.: "Perkins Place." Archaeological site survey form filed by Mariella Squire Hakey, archaeologist, Lyndon State College, 31 Dec 1985.

Vermont Historical Society. (1923–25). Reminiscences by Col. Frye Bailey. *Proceedings.* P. 30.

Washington Papers at the Library of Congress, 1741–1799. Retrieved from http://memory.loc.gov/ammem/gwhtml/gwhome.html.

Watkins, L. W. (1950). *Early New England Potters and their Wares.* Harvard University Press.

Wason, P. K. (1994). *The Archaeology of Rank.* In series: New Studies in Archaeology. Cambridge University Press.

Weber, S. B. (1990). *The History of Greensboro: The First Two Hundred Years.* Greensboro, VT: Greensboro Historical Society.

Wells, F. P. (1902). *History of Newbury, Vermont.* St. Johnsbury, VT: The Caledonian Co.

Wells, T. (1998) Nail Chronology: The Use of Technologically Derived Features. *Historical Archaeology.* 32(2):78–79.

Weaver, E. M. (2013). The Fort Pitt Block House. Stroud, Gloucestershire: The History Press.

White, V. D. (1991). *Genealogical Abstracts of Revolutionary War Pension Files: Index.* Waynesboro, TN: National Historical Publications Co.

White, V. D. (1995). *Index to Revolutionary War Service Records,* Vol. IV: S-Z. Waynesboro, TN: National Historical Publications Co.

Whitelaw, J. (1794). Map of Lake Beautiful (Caspian) Lake, Greensboro, VT.

Whitelaw, J. (1918). Greensboro Proprietors Map. The Secretary of State (ed.), *Index to the Papers of the Surveyors-General. State Papers of Vermont.* Vol. 1:77–78. Rutland, VT: The Tuttle Company.

Wilbur, C. K. (1992). *Home Building and Woodworking in Colonial America.* Guilford, CT: The Globe Pequot Press.

Wilgus, W. (1941). *The Role of Transportation in the Development of Vermont as it was in 1791.* Vermont Department of Highways. Vermont Historical Society Library. (1945).

Wood, F. J. (1919). *The Turnpikes of New England and Evolution of the Same through New England, Virginia, and Maryland.* Boston: Marshall Jones Company.

Index

Made in United States
Orlando, FL
04 November 2026

90201112R00117